TRANSACTIONS

OF THE

AMERICAN PHILOSOPHICAL SOCIETY

HELD AT PHILADELPHIA

FOR PROMOTING USEFUL KNOWLEDGE

VOLUME XXI—NEW SERIES

PART V

Philadelphia:

THE AMERICAN PHILOSOPHICAL SOCIETY

104 SOUTH FIFTH STREET

1908

TRANSACTIONS

OF THE

AMERICAN PHILOSOPHICAL SOCIETY

HELD AT PHILADELPHIA

FOR PROMOTING USEFUL KNOWLEDGE

VOLUME XXI—NEW SERIES

PHILADELPHIA
PUBLISHED BY THE SOCIETY
1908

Press of
The New Era Printing Company
Lancaster, Pa.

CONTENTS OF VOL. XXI.

ARTICLE V.

A SEARCH FOR FLUCTUATIONS IN THE SUN'S THERMAL RADIATION THROUGH THEIR INFLUENCE ON TERRESTRIAL TEMPERATURE.

By Simon Newcomb.

(Read October 4, 1907.)

PREFATORY NOTE.

The purpose of the following study is two-fold. The subject of periodicity in meteorological phenomena, and its relation to the sun, is prominent in scientific literature ; and the author desired to treat it by methods different from the usual ones. He also wishes to submit to the courteous consideration of meteorologists the question whether the methods here developed can not be advantageously used in other branches of their science.

The work has been. carried through under the auspices of the Carnegie Institution, the Trustees of which have enabled the author to avail himself of the necessary appliances, facilities, and computing assistance. Acknowledgment is also due to the U. S. Weather Bureau, the Chief of which has placed at the author's disposal, without restriction, the rich body of material contained in its records, as well as the printed collections in its library ; and to the Director of the *Deutsche Seewarte* of Hamburg for the courteous transmission of unpublished material.

ANALYTICAL TABLE OF CONTENTS.

§ 15. *Search for variations synchronous with the sun's synodic rotation by the method of time-correlation.* When a tendency toward a periodic variation can be expected — choice of San Diego as a station — time correlation through a period of 33 years from 1872 to 1904 ; result, only a suspicion of a periodic tendency, the amplitude being two or three-hundredths of a degree — further illustration of the method from the general ten-day departures — tendency toward a persistence of temperature conditions through periods of more than 40 days. Pp. 375–379.

CHAPTER VI.
DISCUSSION OF RESULTS.

§ 16. *Summary of conclusions.* Actuality of the sun spot fluctuation — uncertain evidence of fluctuations having a shorter period — limitation within which the conclusions are to be interpreted. Pp. 379–381.

§ 17. *Relation between solar radiation and meteorological processes.* The present study limited to thermal radiation — the question whether other emanations producing auroræ and magnetic storms have any appreciable thermal effect — relation between fluctuations of the solar radiation and the fluctuations of temperature hence arising — possible causes of change in the solar radiation — definitive outcome of the investigation. Pp. 381–384.

§ 18. *Comparison with results of Langley's work of 1903.* Pp. 384–387.

INTRODUCTION.

The view that the rate at which the sun radiates thermal energy is or may be variable finds frequent expression in scientific literature. The inference of such variability may be drawn from two sources ; one direct measures with the bolometer, the other, meteorological phenomena, especially variations of temperature at the earth's surface. Many years ago Lockyer pointed out that a cycle corresponding to that of the solar spots was indicated in the agricultural productions of India. A similar cycle has been sought for in the variations of temperature at special places, and in a variety of meteorological phenomena. Brückner has in an elaborate work adduced evidence to show a cycle of about 35 years in meteorological changes generally, those of temperature included. Although the fluctuations here described are not always expressly attributed to the action of the sun, it would be difficult to account for them in any other way than by fluctuations in the sun's radiant energy.

Bigelow's many and long-continued researches on meteorological phenomena, with the view of determining their laws and periods of variations and their relation to the activity of the sun, have also led him to an affirmative conclusion. The best marked period he has sought to establish is one corresponding to the period of the sun's synodic rotation. But the actual conclusions deducible from his work seem to relate to the electric and magnetic effects of the solar activity, rather than to purely thermal effects, which alone are studied in the present work.

Strong evidence on the affirmative side of the question was adduced by Langley, in a discussion of bolometric measures of the sun's radiation in 1902–3, compared with fluctuations in the general terrestrial temperature. During the year 1903 especially, the bolometer showed well-marked periods during which there seemed to be a remarkable diminution of intensity of the sun's radiation. On comparing these fluctuations with those of the temperature in various regions of the globe, derived from the *Dekadenberichte* of the Hamburg Seewarte, a seeming correspondence was shown between the two classes of fluctuations. The relation was exhibited by curves, but was not reduced to the form of an exact numerical relation with a determined probable error.

Notwithstanding the volume of observation and investigation bearing on the subject, and generally supposed to point to the actual existence of fluctuations in the sun's heat, the question cannot be regarded as settled until more precise numerical results than any yet reached are worked out. The drawing of conclusions from any system of direct measures of the sun's radiation, whether made by the bolometer or any other instrument, is subject to the seemingly insurmountable difficulty that the variations in the transparency and temperature of the atmosphere, especially in the higher regions, which may materially affect the measures, cannot be accurately determined. It is equally impossible to determine with precision the varying fraction of the heat which may be intercepted by the atmosphere, and to eliminate the radiation of the matter contained in the atmosphere itself. The uncertainty arising from these ever-varying causes might indeed be reduced indefinitely by comparing simultaneous observations at points so widely separated that no common atmospheric cause could affect the measures at any two stations. But, so far as the writer is aware, no attempt to organize such a series of determinations has yet been made.

On the other hand, when it is proposed to detect fluctuations in the solar radiation by observations of temperature, we meet with the difficulty that the temperature is everywhere subject to fluctuations from local causes, especially the varying aërial circulation, which it is impossible to determine, or to eliminate individually. Hence, in studying the fluctuations of temperature at any one place or in any one region, the problem arises of distinguishing between those due to local causes, and those due to changes in the original source of heat.

The purpose of the present work is to develop and apply the methods best adapted to secure definite results, especially the methods of investigating correlations between irregularly fluctuating quantities. The fundamental principle of this method is the same as that applied by the author long ago in collaboration with E. S. Holden, in discussing the question whether measured variations in the sun's apparent diameter were real; and, more recently, whether there existed any tendency toward unisexuality in families. This method is applicable to fluctuations so irregular that no law, periodic or otherwise, can be detected in their course. Periodicity is to be detected by other methods, involving somewhat different principles, which will also be developed.

In investigating the question it is well to consider in advance the general character of the fluctuations which may be expected. The first question to arise is : assuming that the sun's activity, as determined by terrestrial observations, is subject to a periodic change, what periods are the most likely? The reply to this is that there are only two periods which can be assigned in advance with any plausibility. One is that of the sun-spots ; the other that of the sun's synodic rotation. The latter period would arise if one hemisphere of the sun were occasionally at a higher temperature than the other through two or more successive rotations. We must regard this as highly probable if the solar radiation is subject to any change whatever. It is, in fact, rather unlikely that any cause affecting the temperature of the solar envelope would act at one and the same time over the whole of the photosphere. If a difference in the two hemispheres were permanent, or even if it continued through large fractions of a year, there would be no difficulty in detecting it. As a matter of fact, permanence is scarcely to be expected, and it is in consequence difficult to distinguish between irregular fluctuations and those having this origin.

Granting that some region of the photosphere experienced a rise or fall of temperature which continued through an entire rotation, the effect would be seen in a corresponding fluctuation in the general temperature of the earth. From what is known of motions in the photosphere, it is clearly impossible that two different regions of the solar photosphere at the same latitude and the same altitude can be permanently at different temperatures. But even if the difference in question ordinarily continued only through two or three months, there would be no difficulty in detecting the periodic effect as special regions of the photosphere would successively be brought into view by the sun's rotation. On the other hand, if the inequality of temperature did not ordinarily continue through a single rotation, the effect could not be distinguished from that of irregular fluctuations.

The problem of determining whether there is any period in terrestrial temperature corresponding to that of the solar spots is one of such simplicity that it need not be dwelt upon in the present connection. It will be studied in the course of the present paper.

The really difficult problem is that of detecting with certainty irregular fluctuations in the radiation. The difficulty arises from the fact just mentioned that the fluctuations of temperature are everywhere determined by varying and accidental meteorological causes, especially the motion of large bodies of air from one region to another, and the varying presence of water in its various forms in the atmosphere. Leaving out these disturbing causes it is very natural, when the temperature of a wide region is markedly above or below the normal for a considerable period, to attribute the condition to a change in the amount of heat received from the sun. The question of the reality of this cause admits of an obvious test. A change in the sun's radiation will necessarily affect every part of the earth. If therefore a change of temperature in one region has this cause as a factor we may, accidental causes aside, expect a similar change in every other region. The problem is thus reduced to that of detecting a correlation between the fluctuations of several varying quantities.

Since the ordinary fluctuations of temperature are mainly due to local causes, we may expect the average or general temperature of the entire globe to be sensibly constant if the sun's radiation is invariable. To speak more precisely if, on any one day, it is found that the temperature in every part of the earth is in the general average above or below the normal, we might rationally attribute this result to the sun. We thus see that a very obvious way of testing the constancy of the solar radiation is to determine the deviation of the temperature from the normal on any one day over all points of the globe, and form their mean. The fluctuations of this mean would represent those of the sun's radiation.

It being impossible to extend observations over the entire globe we must accept the results of observations made within regions at which observations of temperature are actually available. But even then it would be an error to conclude that variations in the general mean must be due to the sun or any other common cause. It is not to be expected that the accidental deviations in different regions completely neutralize each other. The question must therefore be open, after we have determined the changes of mean temperature from time to time over the whole globe, whether the mean fluctuations outstanding are purely accidental, or are due to changes in the thermal energy received from the sun. A rigorous method of treating this question will also be developed.

It follows that, in order to reach a well-grounded conclusion, some criterion is necessary to determine whether the changes in the general temperature of the globe are due to changes in the solar radiation, or to accidental terrestrial causes. No criterion which will decide this question in any individual case is possible, but there is a criterion by which the average amount of the cosmical fluctuation, if it be appreciable, can be determined. To show the simplest example of its application let the deviation of the temperature from the normal be observed from day to day and from year to year in two regions of the earth so widely separated that no common purely terrestrial cause can affect the two places at the same time. Then, by the law of probabilities, we should find in the long run that there was no permanent correlation between the fluctuations at the one place and at the other. For example, calling the two regions A and B, if we put into one class all the days on which the temperature in region A is markedly above the normal, and in another class all the days in which it is markedly below normal; and if we take the temperatures in the distant region B for the same two classes of days, then, in the absence of any correlation, we should find the mean temperatures at B to be the same in the two classes. If we found that the mean temperature at B was above the normal when it was above the normal in A, and below it in the contrary case, it would show that there was some common cause affecting the two places. Should the mean temperature in B be entirely independent of that in A it would show that there was no common cause affecting the temperature of the two places and therefore that the fluctuations were not due to changes in the sun's radiation.

By this criterion the existence of either periodic or non-periodic changes can be equally well established, provided that a sufficiently long series of observations is made use of. But it does not enable us to determine the law of change, but only the general fact. When the general form of the law is known, especially when the fluctuations are of definite period, other methods may be applied.

CHAPTER I.

METHODS OF INVESTIGATING FLUCTUATING QUANTITIES.

§ 1. *Fluctuations in a Fixed Period.*

The quantities with which we are concerned in the present paper are in the nature of observed departures from normal or mean values. Such departures may be either results of observation, or they may be derived *a priori* from some theory which is to be tested by observation. Those considered in the present paper are of the first class. We shall take up the general problem of studying fluctuations by considering it in the form suggested by the special problem now before us.

At every place and in every region on the surface of the earth there is for every day a certain mean temperature, best determined by reading the thermometer at a number of equi-distant intervals. These means may be extended through periods of any length, thus giving a series of temperatures extending indefinitely year after year. The temperatures thus observed undergo fluctuations in an annual period, which may be represented either by a Fourier series, or by a smoothed curve extending as nearly as may be through all the observed temperatures. A normal mean temperature for each day throughout the year at any one place may thus be determined from the observations of a number of years—the more the better. Subtracting the normal temperature of each day, or through a period of several days, from the mean temperature actually observed through the same period, we have a certain departure from the normal, due to accidental or systematic causes. To fix the ideas I shall designate the period for which the mean of these departures is taken as a *time-term*, or *term* simply. The data then given by observation comprise the mean departures for a long number of terms, each considered as a unit, and forming so far as possible a continuous series.

The most obvious classification of such departures is into periodic and irregular. In the rigorous mathematical sense a periodic departure is one which always returns to the same value at the end of an interval P of time, called the period. This may be either known or assumed in advance, or regarded as unknown. It cannot, however, be determined as an unknown quantity from conditional equations, because it is impracticable so to introduce it as to give the equations a soluble form. If not regarded as known we have to proceed by the method of trial and error. In this form the question will be whether a certain assumed period P is indicated by observed departures. If the fluctuation had no other term than a purely periodic one as thus defined, its existence could be ascertained by simple inspection. Imagining the fluctuations to be expressed by the ordinates of a curve of which the abscissa is the time, we only have to measure on the axis of abscissas from any arbitrary point, the series of distances P, $2P$, $3P$, etc., to the end of the series. We then take a number of intermediate points and erect at each an ordinate expressing the observed departure. If P is the true period the ordinates would have the same value at all the points distant from each other by a multiple of P. Practically, however, we always have to deal with the case in which other fluctuations than those of period P enter. We thus have accidental deviations superposed upon the periodically recurring departures, which may quite mask them. In this case it is necessary to take the mean value of the observed departure at the several moments P, $2P$, etc., after the initial moment. The mean of all these values would be that corresponding to the initial phase. Taking, as an example, the fluctuations represented in Figure (2), we see that the departure is positive at the beginning of a period.

The method of deciding whether a fluctuation of an assumed period P really exists is this. We divide each period into any convenient number of equal parts by the points 1, 2, 3, etc. We then taket he mean of all the ordinates at the several points 1 ; the mean for the points 2, for the points 3; etc. The several means then show the mean fluctuation during any one period. The absence of any fluctuation in the given period would be shown by these mean values differing from each other only by quantities which might be the result of the accidental deviations.

If the period is unknown, we must discover it tentatively by taking for P the value which gives the best marked mean fluctuation, or the greatest range of value among the mean departures.

In the numerical computation on this principle, after the period is known, or has been discovered, the most general mode of proceeding is that of development in a Fourier series. We take an angle N increasing uniformly with the time at such a rate that it goes through $360°$ in the period P. Then, if we represent the departure at any time by v, we assume it, considered as a function of the time, to be developable in the form

$$v = x_0 + x_1 \cos N + x_2 \cos 2N + \cdots + y_1 \sin N + y_2 \sin 2N + \cdots$$

Regarding x_0, x_1, x_2, $\cdots y_1$, y_2, as unknown quantities, the coefficients of these quantities at each epoch of observation will be the sines or the cosines in the second number of the equation. Substituting for each moment of observation the values of these sines and cosines, and taking the observed departures for v, we shall have a system of equations for determining the unknowns. The solution of these equations by the method of least squares will give the values of the unknowns which best represent the observations.

This method is sometimes employed in meteorology to determine and express the diurnal and annual fluctuations in the temperature. For reasons not necesssary to detail at present, the method of forming the mean values, in the manner first set forth, and then finding the curve which best fits them, is preferable except when, for any reason, all multiples of N above the first are omitted. In this last case the fluctuation will be a purely harmonic one, the coefficients of which can be determined with great facility by equations of condition. An example will be given in investigating the fluctuations in temperature having the sun-spot period.

§ 2. *Irregular Fluctuations Tending Toward a Definite Period, — the Method of Time Correlation.*

There is a class of fluctuations in which the period may be fairly definite, but yet for which the preceding method would give no period whatever. This occurs when we have a superposition of two classes of causes, or two sources of departure, one of which, by itself, would result in a fluctuation in a definite period, while the other is in the nature of perturbations, resulting in disturbances of the phase either continuously or from time to time, and leading to seeming frequent changes in the length of the period. If the preceding or any other method resting on the assumption of unchanging period be applied to this case, the result might be that no period whatever would give a definite fluctuation. In other words a series of departures taken at equi-distant moments would, in the long run, have for their mean either an evanescent value, or a constant value for all phases.

To this class of fluctuations belong the ocean waves. If these are carefully observed we shall generally find in them a tendency in a given state of the weather to follow each other at fairly definite intervals, perhaps at 10, 15 or 20 seconds, according to the distance between the crests. But should we take the mean period, however exact, and record the phase of the wave at any long series of moments separated by exactly this period, we should find no one phase always recurring at the moments thus defined. After a few seemingly regular recurrences of the wave, its height diminishes, perhaps almost to zero, or a fresh series of waves of similar period begins at a different phase from that determined by the preceding waves.

Another case of the same kind is afforded by the swing of a pendulum which is subjected to a continually repeated disturbance, sometimes nearly stopping it, sometimes accelerating it, and sometimes changing the phase of the swing. How frequently soever these disturbances may follow each other, there will always be in the motion of the pendulum a tendency toward its regular period as a function of its length. But it may be impracticable to determine any definite time of oscillation through a long series of observations. In cases like these the perturbations may be so considerable, and follow each other at such short intervals, and the regular mean amplitude of the fluctuation may be so small or variable, that it will be impossible to detect the tendency toward a regular period, except by the application of some special method. To devise a method we must find some criterion for distinguishing between a tendency toward a definite period and complete irregularity.

Any tendency toward a definite period P may be defined in the following way : Let τ be the observed departure at any moment, and τ' the departure at a definite interval P following it. Now if there be really a tendency toward the period P, τ' should differ from τ only by the difference of perturbations, or accidental deviations, which may however be larger than either of the undisturbed departures, and therefore may completely mask the tendency toward equality between τ and τ'. However this may be, the undisturbed departure midway of the period, that is, at the moment $\frac{1}{2}P$, will have the opposite sign τ and τ'. It follows that in the general average, by comparing a large number of departures in triplets, the individual members of which are distant $\frac{1}{2}P$, and calling τ_1 the mean of all the middle departures, the excess of $\tau' - \tau_1$ will in the general average be opposite in sign from τ' itself. If a period be found for which this holds true in the general mean, we have a tendency toward a rhythmical movement in the period P.

The detection of such a period is easy by a method which we may call that of *time-correlation.* The nature of the criterion will be most readily seen by the graphic representations in Figures 1 and 2. Let Figure 1 represent an approximately harmonic fluctuation. If the ordinate at 0 represents the initial variable quantity, there

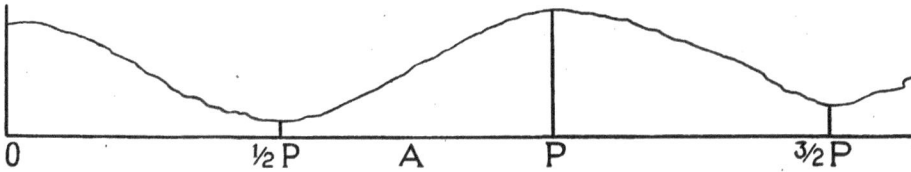

FIG. 1.

will always be a rising phase between the points $\frac{1}{2}P$ and P; say near the point A at $\frac{3}{4}$ of a period from O. If our initial departure is near $\frac{1}{2}P$, then we shall have a descending phase betweeen P and $\frac{3}{2}P$, which is $\frac{3}{4}$ of a period further on.

Now, imagine that the regular fluctuations thus represented have superimposed upon them accidental deviations so large as to mask the harmonic character of the fluctuations. Were these accidental deviations superimposed upon a harmonic motion in a continuous succession of periods, they could be detected by continuing one system of observations through a number of periods, because they would then be eliminated from the mean. But we are supposing a case in which the period is itself disturbed.

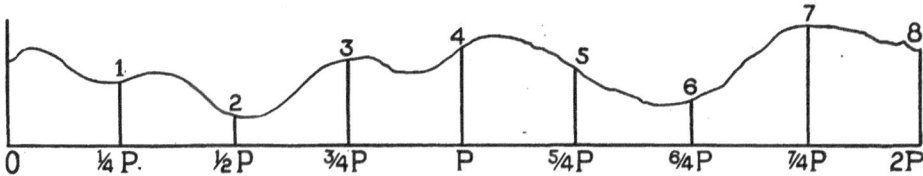

FIG. 2.

What we therefore have to do is to take a number of starting points, numbered 0, 1, 2, etc., and continue the series from each so far as we deem it useful to do so. In these several series the accidental deviations will still be eliminated, ultimately leaving in the general mean a tendency toward the harmonic phase as described.

Such a case is shown in Figure 2. Here there is not evident to the eye any tendency toward an exact period. But a study of the diagram shows that by measuring off equidistant intervals to the points P, $2P$, etc., the departure is, in the general mean positive, while at the middle points of the spaces it is, in the general mean, negative. A criterion is thus offered by which any periodic tendency may be brought out.

We shall now show the method of time-correlation by which not only a period of any length, but any tendency toward a period, may be shown. The period being regarded as entirely unknown, the observed departure from the general mean at any moment may be regarded as due to the periodic term which we seek, with accidental deviations superimposed. Let us put a_0 for the departure at some initial moment; then let us take a series of equi-distant intervals of time, starting from the initial one, and let us put

$$a_1, a_2, a_3, \cdots, a_n$$

the deviations at the ends of the intervals t, $2t$, $3t$, $\cdots nt$. If there is any tendency toward a rhythmical motion in these departures, having a period greater than $2t$ but less than nt, then, in the general average, assuming a_0 to be positive, we should find first a diminution and then an increase in the series of a's; that is, the curve representing the departures would be convex to the axis of abscissas.

Since one departure may be taken for the initial one as well as another, we may repeat this process with a_1, a_2, etc., as the initial departures, carrying the products in each case to the requisite number of terms. We shall thus have a series of products which may be continued as far as the series of observed departures extends. Taking as an example $n = 5$, the arrangement is the following:

$$
\begin{array}{cccccc}
a_0^2 & a_0 a_1 & a_0 a_2 & \cdots & a_0 a_5 \\
a_1^2 & a_1 a_2 & a_1 a_3 & \cdots & a_1 a_6 \\
a_2^2 & a_2 a_3 & a_2 a_4 & \cdots & a_2 a_7 \\
\vdots & \vdots & \vdots & & \vdots \\
\end{array}
$$

Sums: $[aa]$ $[aa_1]$ $[aa_2]$ $\cdots$ $[aa_5]$

This arrangement suggests the solution by least squares of a problem which may be put into the following form. Starting as before with the initial departure a_0, if the fluctuation be a purely harmonic one, the departure at the end of half a period would always be $- a_0$, that at the end of a period $+ a_0$, etc. In general the departure at any time t will be of the form $a_0 + x$, x being a periodic function of t. Consequently the actual deviations a_1, a_2, etc., will be of the form

$$a_1 = a_0 x_1 \pm e_1; \quad a_2 = a_0 x_2 \pm e_2; \quad \text{etc.}$$

e_1, e_2, etc., being the purely accidental parts of the deviations. The problem thus resolves itself into determining a series of factors x_1, x_2, x_3, etc., by means of the conditional equations

$$a_1 = a_0 x_1; \quad a_2 = a_0 x_2; \quad \text{etc.}$$

These may be combined by the method of least squares.   The normal equation is

$$[a_0a_0]x_i = [a_0a_i]$$

from which x_i is at once found.   Thus, putting $i = 1, 2, 3$, etc., we shall have a series of quantities

$$x_1, x_2, x_3, \cdots, x_n$$

of which numerical values may be determined from the equations.   A tendency toward a rhythmical deviation of the kind we are in search of will be shown by an increasing value of x at the time corresponding most nearly to the completion of the period.   If there is no tendency toward any period between the limits $2t$ and nt the series of x's will converge in the general mean toward the value zero.

§ 3. *Treatment of Fluctuations without Discernible Period.*

The method developed in the two preceding sections is applicable to a single series of observations of fluctuating quantities of any kind, and will enable us to determine any periodic tendency in them.   We have now to consider the case in which the periodicity is not discernible.   In this case results are to be derived by comparisons of different series of observations made simultaneously at different places.   Our treatment will be that of the special case of departures in temperature; but the method may of course be applicable in the wider field of fluctuating quantities in general.

We know that deviations of the temperature from the normal are of constant occurrence at every point of the globe.   We also know that these are due, in great part at least, to local causes, especially the motion of the air from region to region, and the varying effects of cloud and moisture.   But they may also be due in part to changes in the sun's radiation of heat, or other general causes.   The question is what evidence can be found to indicate the action of a general cause affecting the whole earth simultaneously.   It is plain enough that observations at one place, no matter how long continued, would never enable us to distinguish between fluctuations of temperature due to local causes and to the sun.   But by comparing simultaneous observations in regions of the earth so widely separated that the same local causes could not have influenced the temperatures in both regions, it is possible to determine, approximately at least, by a statistical method which we shall now develop, what part the sun or other general cause may play in the fluctuations.

The data for our problems are the simultaneous departures of temperature from the normal, in a number n of regions, through a series of terms of equal length in time, this length being chosen so as to best meet the requirements of the problem.   Let us put

$$v_1 \quad v_2 \quad v_3 \quad v_n$$

the n simultaneous departures of the temperature from the normal in the n regions for any one term.

Considering the problem thus suggested as that of determining the normal departure of a world temperature, produced by any cause affecting the world earth, such as a change in the sun's radiation, the obvious method of determining this world deviation is to take the mean of all the separate departures v_i, observed in various regions. Let us then put τ, for the apparent mean departure of the world temperature from the general normal. ' This appparent departure is determined by the equation

$$n\tau = v_1 + v_2 + v_3 + \cdots + v_n = \Sigma_i v_i \tag{1}$$

or

$$\tau = \frac{\Sigma_i v_i}{n} \qquad\qquad (i = 1, 2, \cdots, n)$$

Before taking up the question of a cosmical cause affecting the world-temperature, let us consider the problem as that of determining the probable error of the departure of the world-temperature from the normal. To do this we subtract τ from the individual deviations, v_i. We then have a series of residuals, u_1, u_2, etc.

$$u_1 = v_1 - \tau, \; u_2 = v_2 - \tau, \; \cdots, \; u_n = v_n - \tau$$

Following the method of least squares let us form the squares of these residuals

$$u_1^2 = v_1^2 - 2\tau v_1 + \tau^2$$
$$u_2^2 = v_2^2 - 2\tau v_2 + \tau^2$$
$$u_n^2 = v_n^2 - 2\tau v_n + \tau^2$$

Putting ϵ for the mean deviation and summing these residuals we shall have by the theory of least squares the probable equation

$$(n - 1)\epsilon^2 = \Sigma_i u_i^2 = \Sigma_i v_i^2 - n\tau^2 \tag{2}$$

Conceive now that we determine the deviation of the world-temperature in this way through a number of time-terms, arriving at a series of values of τ, each having its mean error ϵ_r. It is clear that the value of the mean error should not be determined separately for each term from the discordances for that term alone, but from the residuals throughout the whole period. If r be the entire number of time-terms the number of these residuals will be nr. We represent by Σ_j a summation through the r terms, and by $\Sigma_{i,j}$ a summation of all the nr values. Then, by adding the r equations of the form (2) we have the probable equation

$$r(n - 1)\epsilon_r^2 = \Sigma_{i,j} v^2 - n\Sigma_j \tau^2 \tag{3}$$

Also, by squaring the equation (1) and adding the r squared equations we find

$$n^2 \Sigma_i \tau^2 = \Sigma^2_{ij} v^2 + 2\Sigma vv'$$

where $\Sigma vv'$ represents the sum of the $rn(n-1)$ products of each two departures in every term. If these departures v are purely accidental deviations from means the ratio of $\Sigma vv'$ to Σv^2 will tend toward zero as the number of terms is indefinitely increased.

Dropping them we find the condition

$$n^2 \Sigma_j \tau^2 = \Sigma_{i,j} v^2$$

Hence, if we put,

$$\Delta = n^2 \Sigma_j \tau^2 - \Sigma_{i,j} v^2 \qquad (4)$$

the criterion for the independence of each v from the others will be

$$\Delta = 0 \qquad (5)$$

If this equation is not satisfied within the probable limits of the accumulated accidental errors, it will show that the hypothesis of the complete independence of the temperatures of the different regions is not established, and that there is some correlation between them. This may arise from any common cause affecting the temperature at two or more of the stations. Let us suppose a varying cosmical cause affecting the entire earth, the result of which is to raise the world-temperature during any one term by an amount τ_0. Each observed departure will then be made up of two parts :—

(1), the common departure τ_0 for the whole world ;

(2), an accidental local deviation peculiar to the region. We shall then have, as the value of each individual departure in any region during any one term

$$v_i = \tau_0 + v_i' \qquad (6)$$

v' being the purely accidental deviation, whose mean value is ϵ.

Form the sum of the squares of the equations (6) for the n values of the v_i for any one term

$$\Sigma_i v_i^2 = n\tau_0^2 + 2\Sigma_i \tau_0 v_i' + \Sigma_i v_i'^2 \qquad (7)$$

The mean value of v' being the same as that of ϵ, and each value of v' being independent of τ_0, we have the probable equation

$$\Sigma_i \tau_0 v_i' = 0$$

Summing the equation (7) for the r time-terms and putting ϵ^2 for the mean v'^2 we have

$$\Sigma_{i,j} v^2 = n\Sigma_j \tau_0^2 + nr\epsilon^2 \qquad (8)$$

Now let us treat the mean departure τ in the same way. We put e, the mean of the purely accidental part of τ. Then in each time-term,

$$\tau = \tau_0 + e$$

Squaring and summing the r values of τ we have

$$\Sigma_j \tau^2 = \Sigma_j \tau_0^2 + 2\,\Sigma e \tau_0 + \Sigma_j e^2$$

For the same reason as in the individual deviations we have

$$\text{Probable } \Sigma e \tau_0 = 0$$

$$\text{Probable } \quad e^2 = \frac{\epsilon^2}{n}$$

and thus the equation becomes

$$\Sigma_j \tau^2 = \Sigma_j \tau_0^2 + \frac{r\epsilon^2}{n}$$

Eliminating ϵ^2 between this equation and (8) we find by using (4)

$$n(n-1)\Sigma_j \tau_0^2 = n^2 \Sigma_j \tau^2 - \Sigma_{ij} v^2 = \Delta \tag{9}$$

The second member of this equation is computed by summing the squares of all the τ's, which are r in number, and also the squares of all the nr individual departures. Having thus found r values of Δ, the sum of which we shall call Δ simply, the probable mean world-deviation τ_0 is given by the equations

$$nr(n-1)\tau_0^2 = \Delta$$

$$\text{Probable mean } \tau_0^2 = \frac{\Delta}{nr(n-1)} \tag{10}$$

When several periods, for which the number of regions was unequal, are to be combined, the final equation for τ_0^2 should be put into the form

$$\Sigma rn(n-1)\tau_0^2 = \Sigma \Delta$$

This value of τ_0^2 will be subject to a probable error arising from the probable accumulation of accidental deviations in the sum of all the quantities which form it. Our conclusions as to its value must depend upon how far its actual value exceeds this probable accidental deviation. If within the limits of probable deviation, we must consider that the evidence is against its having any determinable value. The probability of its having a real value increases with its magnitude as compared with the probability of the accidental value.

It may happen that $\Sigma \Delta$ comes out negative. This would signify that, instead of the simultaneous temperatures in the different regions being independent, or affected by a common cosmical cause, one region on the average becomes hotter or colder at the expense of another. In other words the conclusion would be that when the temperature was above the normal in one region, it was more likely than not to be below it in other regions, and *vice versa*. Thus the conclusion as to a positive correlation,— no correlation or a negative correlation — depends upon whether Δ is positive, evanescent or negative.

§ 4. *Case when Different Weights are Assigned to Different Regions.*

For the sake of simplicity we have developed the preceding method on the assumption that in determining the general departure τ the different stations or regions are all entitled to the same weight. But if the accidental deviations are smaller at some stations than at others it is clear that the observations at such stations will be of greater weight for the detection of cosmical causes. We should therefore assign weights to the several stations determined by the usual methods. Let these weights be

$$w_1, w_2, \cdots, w_n \tag{11}$$

and let us call W their sum. The preceding equations will then be replaced by the following:

Instead of using (1) for determining τ we use the equation

$$W\tau = w_1 v_1 + w_2 v_2 + \cdots + w_n v_n = \Sigma_i w_i v_i \tag{12}$$

Let us put ϵ_i for the mean accidental deviation of v_i and ϵ_τ for that of τ. The mean deviation of any one product $w_i v_i$ is then $w_i \epsilon_i$ and the squared mean deviation of the sum of all these products for any one term, if uncorrelated, is

$$\Sigma w_i^2 \epsilon_i^2$$

The mean ϵ_τ should in this case satisfy the equation

$$W^2 \epsilon_\tau^2 = w_1^2 \epsilon_1^2 + w_2^2 \epsilon_2^2 + \cdots + w_n^2 \epsilon_n^2 \tag{15}$$

If the observed deviations v are wholly in the nature of accidental deviations from a mean value, we may take for each ϵ_i^2 the mean of all the v_i^2; and τ being then a purely accidental deviation of the mean, we should have the probable equation

$$\epsilon_\tau^2 = \text{mean } \tau^2 .$$

The criterion for deciding whether the deviations are purely accidental may therefore be written in the form $\Delta = 0$, where for any one time-term

$$\Delta_i = W^2\tau^2 - (w_1^2 v_1^2 + w_2^2 v_2^2 + \cdots + w_n^2 v_n^2)$$

There being r time-terms in all, each will give a value of Δ_i the sum of which we call Δ simply.

Summing all r of these probable relations the criterion will become

$$\Delta = \Sigma_j W^2\tau^2 - \Sigma_j\Sigma_i w_i^2 v_i^2 = 0 \qquad (j = 1, 2, \cdots, r) \quad (16)$$

If the value of Δ comes out too large to be regarded as the accumulated effect of chance deviations, we must, as before, find a mean deviation τ_0, common to all the stations for each separate term of observation, which will reduce the second member to the value Δ. We do this by the same process as that when the weights are taken as equal. We have, as before, the probable equations

$$\Sigma\tau^2 = \Sigma\epsilon_0^2 + \Sigma\tau_0^2$$
$$\Sigma_i v_i^2 = \Sigma_i \epsilon_0^2 + \Sigma\tau_0^2$$

Substituting these values in (16) the terms in ϵ^2 all drop out by virtue of the relation (15), and we have left the probable equation

$$\Sigma_j(W^2 - \Sigma_i w^2)\tau_0^2 = \Delta \qquad (17)$$

which determines a probable mean value of τ_0^2, and therefore of τ_0 on the same principles as when the weights are equal.

§ 5. *Comparison of Regions when Taken by Pairs.*

When only two regions are compared the process of § 3 may be simplified. Calling v and v' the observed departures when only two regions are considered we shall have

$$2\tau = v + v'$$

for each term of observation. Hence

$$2\tau^2 - \tfrac{1}{2}(v^2 + v'^2) = vv'$$

Summing for all r terms, as before

$$2\Sigma_j\tau^2 - \tfrac{1}{2}\Sigma_j(v^2 + v'^2) = \Sigma_j vv'$$

Thus, putting $n = 2$ in (9) and (10) we find for each time-term the simple expression

$$\text{Mean } \tau_0^2 = \frac{\Sigma vv'}{r} = \text{Mean } vv' \qquad (18)$$

which is much simpler in this case than the formula (10).

We may, if we choose, reduce the results for any number of regions in the same way by taking the regions in pairs. By squaring (1) we have, for any one term of observation,

$$n^2\tau^2 = \Sigma_i v_i^2 + 2\Sigma_{i,k} v_i v_k \tag{19}$$

in which each individual product $v_i v_k$ is formed from each pair of the individual v's for the time-term, so that we have $n(n-1)$ products $v_i v_k$ for each of the r time-terms.

Summing the series for all the time-terms during which n remains the same, we have

$$n^2\Sigma_j\tau^2 = \Sigma_{i,j} v^2 + 2\Sigma_{i,j,k} v_i v_k \tag{20}$$

Combining this with (9) we have

$$n(n-1)\Sigma_j\tau_0^2 = 2\Sigma_{i,j,k} v_i v_k$$

Taking τ_0 to represent the mean value of the cosmical fluctuation through r terms, we have

$$\Sigma_j\tau_0^2 = r\tau_0^2 \tag{21}$$

Also,

$$\Delta = 2\Sigma^3 vv'$$

where, for brevity, we put Σ^3 for the triple summation of the products. We are thus enabled, when we so desire, to compute Δ, and hence the value of τ_0^2, for each time-term and each pair of stations taken separately. The final mean of τ_0^2 which we thus derive instead of (10) is

$$\text{Mean } \tau_0^2 = \frac{2\Sigma^3 vv'}{n(n-1)} \tag{22}$$

The number of combinations of n stations being $[n(n-1)]/2$, this is equivalent to

$$\text{Mean } \tau_0^2 = \text{Mean } vv' \tag{23}$$

which may be found by summing (18) for the pairs of stations and all the time-terms. For considerable values of n this equation is more laborious in use than (10) or (17), but it has the advantage of showing whether a correlation among the departures of temperature exists for all the stations, or is confined to a limited number of stations.

The preceding value of τ_0^2 has been derived for the sake of simplicity, as if the weights were all equal. When the pairs of stations are all considered individually, no difference of assigned weights will affect the resulting individual value of τ_0^2. But if we combine the $[n(n-1)]/2$ individual values thus derived, we must assign them their proper weights. These we find by dealing with (16) in the same way that we have dealt with $n^2\tau^2$ when the weights were each 1. By squaring (12) and summing for the r time-terms, we find

$$\Sigma_i\Sigma_j w_i^2 v_i^2 = \Sigma_j W^2\tau^2 - 2\Sigma_{i,k,j} w_i w_k v_i v_k \tag{24}$$

This, substituted in (16), gives

$$\Delta = 2\Sigma_{i, k, j} w_i w_k v_i v_k \tag{25}$$

With equal weights we have, from any one pair of stations, $(n = 2)$

$$\Delta = 2\Sigma_j vv' \tag{26}$$

It follows that if we put $\Delta_{i, k}$ the special value of Δ found for any pair of stations without regard to weights, the final value for use in (17), when the weights are taken account of is

$$\Delta = \Sigma_{i, k} w_i w_k \Delta_{i, k} \tag{27}$$

and we shall then have

$$\text{Probable mean } \tau_0^2 = \frac{\Sigma_{i, k} w_i w_k \Delta_{i, k}}{\Sigma W^2 - \Sigma_{i, j} w^2} \tag{28}$$

CHAPTER II.

REVIEW OF DATA AND PROCESSES.

§ 6. *Choice and Combination of Material.*

From the preceding exposition of the general method applied it will be seen that, since our result is based on systematic or accidental departures alone, and not on absolute temperatures, our main requirement is long series of observations of temperature, at widely separated points of the earth's surface, made and reduced on a plan which should be uniform for each point, but might vary to any extent from one point to another. A single observation of temperature on each day would suffice in the long run, provided it were made at the same hour throughout. Of course a better result is reached from a number of daily observations at given hours; but this is less essential than uniformity of system at each separate station.

In planning the work it was hoped that the much-criticised labor spent in accumulating meteorological observations might be found not so ill-directed as is sometimes thought. Unvaried routine, even if unintelligent, in the method of making and publishing the observations would be an advantage in a case where errors and defects in the instruments and methods are unimportant for the result, so long as they remained unchanged. But, when the actual material was sought out and examined, disappointment was nearly everywhere the result. Outside a few government establishments supported by civilized nations or other permanent organizations, diversity instead of uniformity was found to prevail, — even unintelligent adherence to any routine system of making, reducing, and publishing the observations being rare. The amount of available material was also diminished by the fact that a very important part of the best-planned meteorological observations are made only to determine the climatology of the region, and are abandoned when this requirement is satisfied. The importance of supplying in a satisfactory way this want of uniformity and continuity has given a certain disjointed character to the material used in the present investigation. With this preliminary remark we pass to the selection of the actual material.

Since the effect of any change in the daily amount of energy radiated by the sun will be more strongly felt in those regions most exposed to that radiation, it follows that tropical stations should have the preference over those of high latitudes. At the same time, the longer the period through which a set of observations extends the less the importance of this preference. I have therefore not made use of observations in the northern countries of Europe in comparing and observing monthly and ten-day means; but have utilized a wider range of annual means. No precise limits as to latitude have been set in any one case, the choice necessarily depending on general availability.

Deviations of temperature have less weight the wider the range of accidental variation from day to day. It was therefore deemed advisable to omit regions where the temperature fluctuated rapidly. But this requirement also was relaxed in case of terms of long period, because the purely accidental effects would be more and more diminished as longer periods were taken.

In selecting records to be used we must distinguish the essential from the non-essential features. As the object is not to determine the actual mean temperature in the several regions, but fluctuations only, it is nearly indifferent how the daily means are derived. The mean temperature for the whole twenty-four hours is preferable to a single observation at one and the same hour only because the purely accidental deviation will then be smaller. This actual mean is also preferable to the mean of the maximum and minimum temperatures, but the advantage in this case is not sufficiently well marked to justify a great expenditure of labor to secure it. What is essential is that a uniform system of observed temperatures should extend through a sufficient number of years to enable a table of normal temperatures for each month or each decade of the year to be formed. But it is not necessary that even this table should be one entitled to great weight. In fact without any normal standard, the mean deviations from day to day, or from period to period, would be entitled to some weight. While some pains have been taken to construct a table of normal temperatures for several of the stations, this part of the work has not been regarded as definitive, and is not published in this paper.

From the nature of our method, as developed in the preceding chapter, our first step must be to divide the surface of the earth into regions, within each of which the accidental changes of temperature may be supposed independent of those in every other region. Having done this, we are not confined to a single observing station in each region. In fact the more observing stations used in each, and the more widely they are separated, the greater will be the weight to which the mean result for the region is entitled.

We shall now review the material made use of, and the method of handling it, so far as seems necessary to enable critical investigators to examine and test the processes in detail, and to form a judgment as to the reliability of the result. An entirely systematic statement of the plans and methods is difficult from the fact that they had to be changed in detail from time to time as the work progressed, owing to the unexpected cases of incompleteness and other imperfections which showed themselves here and there as the compilation went on. Lack of uniformity in treatment has also arisen from the discovery from time to time of new material which it was thought advisable to use in the work. Moreover a certain perfection of method originally aimed at, involving a rigorous reduction to a 24-hour mean for every day, was found impracticable, and such means as chanced to be available had very generally to be used. The effect of this drawback upon the results of the work itself is practically quite unimportant; but it prevents the material made use of from having the completeness and certainty that it otherwise might have claimed as a basis for more extended meteorological researches.

It may be added that the conclusions reached in the research can be judged without any reference to the original materials on which the work is based; but, as already intimated, a knowledge of this material will not only facilitate the judgment of any details but will be of assistance to any one desiring to review and extend the work. The following are the sources from which the data were mostly derived.

Records of the United States Weather Bureau. — The original plan was to choose a number of widely separated stations in the United States and, from the manuscript records of the Bureau, to reduce the recorded mean temperature of each day to the rigorous 24-hour mean, and then obtain a daily deviation from the normal during the entire period. But the discussion of the entire 35 years of records on this plan was found to be too laborious, especially as the hours and system of observation had been changed twice during the history of the bureau. It was therefore deemed sufficient to take as the standard temperature for each day the mean of the maximum and minimum temperatures. This was, for the most part, reduced to the 24-hour mean when data for doing so could be readily found.

The Argentine Republic. — The main source for this region has been the publications of the *Officina Meteorologica Argentina*. I am also indebted to Dr. Davis, Director of the Meteorological Office, for the communication of observations additional to those found in the published volumes. The number of stations used in these years was different in different years. Generally six or eight were available.

Havana: Observaciones Magneticas y Meteorologicas del Real Collegio de Belen de la Compania de Jesus en la Habana. Habana, Imprenta del Avisador Commercial Aucargura 30, 1890.

Jamaica: Temperatures in Kingston, Jamaica. Jamaica; Government Printing Office. Doc. No. 275.

Mauritius: Meteorological Observations taken during a number of years, and published annually as a Mauritius blue book.

Bombay: Magnetic and Meteorological Observations at the Government Observatory, Bombay. Bombay, printed at the Government Central Press, 1895.

Batavia: Observations made at the Royal Magnetical and Meteorological Observatory, Batavia, 1900. Here only one station is available and the deviations as will be seen from the table are larger in the mean than in the case of any others that have been included. They received therefore only the weight 1.

Ceylon: Administrative Reports on Meteorology. No general title in detail. These publications contain monthly and annual means of observations at a number of stations on the island. The deviations used here and elsewhere are the means of generally six or more stations in various parts of the island.

Australia: The sources for these observations are the annual publications of the Adelaide Observatory by Sir Charles Todd. The means given are generally those at five or six different stations.

Madras: Results of the meteorological observations made at the Government Observatory at Madras, — 1861–1890. Madras, 1892. Also other volumes partly without title and partly bearing a similar title.

Manila: Census of the Philippine Islands, 1903, Bulletin 2. The Climate of the Philippines by Rev. Jose Algue, Director of the Philippine Weather Bureau, published by the Census Office, Washington, 1904.

Apia: The *Deutsche Überseeische Meteorologische Beobachtungen* contains meteorological observations at a number of coast and island stations; but, for the most part, the observations are not pursued continuously through a sufficient period to be well adapted to the present work. The best station for our purpose proved to be Apia where the record is nearly complete since 1890. The unpublished results up to 1904 were courteously communicated by the Director of the *Deutsche Seewarte* at Hamburg.

In equability and uniformity of temperature this station not only leads every other on the list but every region. If therefore internal consistency had been the sole guide in assigning weights, it would be entitled to greater weight than any other two stations. But there is always a possibility at any one station of varying systematic errors from one cause or another. Hence, it has received no greater weight than the best of the remaining stations.

§ 7. *Arrangement of the work.*

Owing to the complexity of the conditions which have determined the final form of the work, the task of studying its processes will be facilitated by a condensed statement of its arrangements. The main features to be borne in mind are the following :

Firstly, as regards geographic distribution ; that portion of the earth best available for the purpose is divided into regions within each of which the fluctuations of temperature are *prima facie* independent of those in other regions. The question whether this independence is real is regarded as open to question and therefore has been investigated in special cases where a correlation is possible.

Then, within each region as many stations of observation as practicable are to be selected in order that the accidental fluctuations of the regions may be reduced. Frequently there is but one station in a region.

Secondly, as to the time; the whole period included in each special branch of the discussion is divided up into time-terms. The time-terms actually used are three, (1) the year; (2) the calendar month; (3) the decade.

Were the work ideally complete in every particular, we should logically begin with the decade, then pass to the month, and then to the yearly terms, because this is the order in which the observations are made and the work has to be done. But, for reasons not necessary to set forth at length, the different series of time-terms are presented in reverse order, beginning with the year.

The material used is different for the three classes of terms. In discussing the ten-day terms it was found that, quite apart from the labor of forming ten-day means, the available material in the form of daily observations was comparatively limited. But monthly and annual means are found in so many publications that the data available for this branch of the research is great. This additional wealth of material has permitted the use of a much greater number of regions than are available for the ten-day means.

CHAPTER III.

DISCUSSION OF ANNUAL DEVIATIONS OF TEMPERATURE.

§ 8. *Fluctuations Having the Period of the Sun-spots.*

Proceeding according to the plan mapped out, our first step will be to determine the fluctuations in temperature corresponding to the 11-year inequality in the solar spots. This periodic change in the amount of solar spottedness indicates that a change of some sort is going on in the sun ; and if the radiation of the latter is subject to any periodic change, we must expect this to be one of the principal periods. Two methods of investigation are open to us, which would be identical if the variation in the spotted- ness were a rigorously harmonic fluctuation in a fixed period. One is to take the degree of spottedness from time to time as the term of comparison ; the other is to assume a period in the general terrestrial temperature exactly equal to the mean period of the spots, and determine the coefficients of the fluctuation so as to best satisfy the observations. The second method seems preferable because we have some reason to suppose that the degree of spottedness is a secondary rather than a primary phe- nomenon. The writer showed in his paper on the period of the solar spots that the irregularities in the period of the observed phenomenon tended to compensate them- selves, in the course of time returning to an original primordial period. This was especially shown by the fact that about 1760–90 the epochs of maximum and mini- mum were accelerated for several years, but afterward returned to the original places in the period. That is to say we have in the spots a fairly exact period subject to fluctuations on one side and on the other. Now the change in radiation is as likely to follow the rigorous period as to follow the apparent phenomena of spots.

The irregular and fragmentary character of our data affords another reason for taking as the basis of our work the hypothesis of a period of 11.13 years simply. If we had at our disposal a uniform and homogeneous system of observations in various widely separated regions, extending through a long period, either method could be applied with equal facility. But the fragmentary character of the actual data would render weak a comparison of the temperature during a period of such great bespotted- ness as that of 1870–71 with that of the year 1900, during which there were very few spots.

The most exhaustive attempt with which I am acquainted to discover the relation between the solar spottedness and the terrestrial temperature is that of Köppen.* The material made use of comprises mean fluctuations of temperature in various regions of the globe, from 1767 to 1877. The regions were classified according to their lati- tude as tropical, sub-tropical, warm, temperate, etc. The general conclusion was that the temperature of the tropical regions was lower by about $0°.73$ C. near the time of maximum sun spots than near the time of minimum. It is known that the spots radiate less heat in proportion to their surface than does the photosphere, and the general nature of this result is the same as if the temperature per unit area of the non-spotted photosphere were invariable, so that the total radiation was diminished

* *Zeitschrift der Oesterreichen Gesellschaft für Meteorologie*, VIII **Band**.

by the spots. The fluctuation of terrestrial temperature was shown to be the greatest in the equatorial regions, and to diminish progressively as the latitude increased to north or south. There were also indications of a non-correspondence between the epochs of maximum and minimum temperatures, and the minimum and maximum of spottedness, but the determination of the difference must be considered as weak, in view of the uncertainty of the data and the minuteness of the fluctuation.

The writer proposes to reinvestigate this question, using both Köppen's data and more recent observations, in order to apply the more rigorous method of equations of condition. We assume only that the mean temperature at the earth's surface fluctuates harmonically in a period of 11.13 years. This hypothesis may be represented in the general form

$$\Delta\tau = x \cos \mu t + y \sin \mu t + z \qquad (29)$$

where μ is to be so taken that the angle μt shall go through 360° in the given period. Taking the year as the unit of time this gives

$$\mu = 32°.35$$

The epoch from which t is measured is quite arbitrary, because when, after deriving x and y from observations, we reduce the expression to a monomial

$$\Delta\tau = \rho \sin (\mu t + c)$$

the value of $\mu t + c$ for a given moment of time will be the same, whatever the chosen epoch for $t = 0$.

Putting, for brevity,

$$a = \cos \mu t \,; \; b = \sin \mu t$$

each observed deviation of temperature, $\Delta\tau = r$, will give the equation of condition

$$ax + by + z = r$$

These conditional equations being treated by the method of least squares we shall have the normal equations

$$[aa]x + [ab]y + [ac]z = [an]$$
$$[ab]x + [bb]y + [bc]z = [bn]$$
$$[ac]x + [bc]y + [cc]z = [cn]$$

Having found x and y from these equations we may substitute them in (15), and reduce the trigonometric terms to a monomial by computing ρ and c from

$$\rho \cos c = x$$
$$\rho \sin c = -y$$

The harmonic fluctuations of which we are in search will then be

$$\Delta\tau = \rho \cos (\mu t + c)$$

In the actual investigation I have taken the epoch 1844.6 as that from which t is counted. This epoch corresponds to a sun-spot minimum; but this is unimportant at the present moment. From this starting point the value of the angle of μt was taken for the middle of each year, and its sine and cosine, with their squares and products, were computed with results shown in the following table:

TABLE I.

Coefficients for Detecting Fluctuations Having the Sunspot Period.

Year	a	b	a^2	b^2	ab	Year	a	b	a^2	b^2	ab
1820	+0.5	−0.9	0.25	0.81	−0.4	1865	+0.7	−0.7	0.49	0.49	−0.5
21	+0.9	−0.5	0.81	0.25	−0.4	66	+1.0	−0.2	1.00	0.04	−0.2
22	+1.0	0.0	1.00	0.00	+0.1	67	+0.9	+0.4	0.81	0.16	+0.3
23	+0.8	+0.6	0.64	0.36	+0.5	68	+0.6	+0.8	0.36	0.64	+0.5
24	+0.4	+0.9	0.16	0.81	+0.3	69	+0.1	+1.0	0.01	1.00	+0.1
1825	−0.2	+1.0	0.04	1.00	−0.2	1870	−0.5	+0.9	0.25	0.81	−0.4
26	−0.7	+0.7	0.49	0.49	−0.5	71	−0.9	+0.5	0.81	0.25	−0.4
27	−1.0	+0.3	1.00	0.09	−0.2	72	−1.0	0.0	1.00	0.00	+0.0
28	−1.0	−0.3	1.00	0.09	+0.3	73	−0.8	−0.6	0.64	0.36	+0.4
29	−0.6	−0.8	0.36	0.64	+0.5	74	−0.4	−0.9	0.16	0.81	+0.4
1830	−0.1	−1.0	0.01	1.00	+0.1	1875	+0.2	−1.0	0.04	1.00	−0.2
31	+0.4	−0.9	0.16	0.81	−0.4	76	+0.7	−0.7	0.49	0.49	−0.5
32	+0.8	−0.5	0.64	0.25	−0.4	77	+1.0	−0.3	1.00	0.09	−0.3
33	+1.0	0.0	1.00	0.00	0.0	78	+1.0	+0.3	1.00	0.09	+0.3
34	+0.8	+0.5	0.64	0.25	+0.4	79	+0.7	+0.8	0.49	0.64	+0.6
1835	+0.4	+0.9	0.16	0.81	+0.4	1880	+0.1	+1.0	0.01	1.00	+0.1
36	−0.1	+1.0	0.01	1.00	−0.1	81	−0.4	+0.9	0.16	0.81	−0.4
37	−0.6	+0.8	0.36	0.64	−0.5	82	−0.8	+0.6	0.64	0.36	−0.5
38	−1.0	+0.3	1.00	0.09	−0.3	83	−1.0	0.0	1.00	0.00	0.0
39	−1.0	−0.3	1.00	0.09	+0.3	84	−0.9	−0.5	0.81	0.25	0.4
1840	−0.7	−0.7	0.49	0.49	+0.5	1885	−0.4	−0.9	0.16	0.81	+0.4
41	−0.2	−1.0	0.04	1.00	+0.2	86	+0.1	−1.0	0.01	1.00	−0.1
42	+0.4	−0.9	0.16	0.81	−0.3	87	+0.6	−0.8	0.36	0.64	−0.5
43	+0.8	−0.6	0.64	0.36	−0.5	88	+0.9	−0.3	0.81	0.09	−0.3
44	+1.0	0.0	1.00	0.00	−0.1	89	+1.0	+0.2	1.00	0.04	+0.2
1845	+0.9	+0.5	0.81	0.25	+0.4	1890	+0.7	+0.7	0.49	0.49	+0.5
46	+0.5	+0.9	0.25	0.81	+0.4	91	+0.2	+1.0	0.04	1.00	+0.2
47	−0.1	+1.0	0.01	1.00	−0.1	92	−0.3	+0.9	0.09	0.81	−0.3
48	−0.6	+0.8	0.36	0.64	−0.5	93	−0.8	+0.6	0.64	0.36	−0.5
49	−0.9	+0.4	0.81	0.16	−0.3	94	−1.0	+0.1	1.00	0.01	−0.1
1850	−1.0	−0.2	1.00	0.04	+0.2	1895	−0.9	−0.4	0.81	0.16	+0.4
51	−0.7	−0.7	0.49	0.49	+0.5	96	−0.5	−0.9	0.25	0.81	+0.4
52	−0.2	−1.0	0.04	1.00	+0.2	97	0.0	−1.0	0.00	1.00	0.0
53	+0.3	−1.0	0.09	1.00	−0.3	98	+0.5	−0.8	0.25	0.64	−0.4
54	+0.8	−0.6	0.64	0.36	−0.5	1899	+0.9	−0.4	0.81	0.16	−0.4
1855	+1.0	−0.1	1.00	0.01	−0.1	1900	+1.0	+0.1	1.00	0.01	+0.1
56	+0.9	+0.4	0.81	0.16	+0.4	01	+0.8	+0.7	0.64	0.49	+0.6
57	+0.5	+0.8	0.25	0.64	+0.4	02	+0.3	+1.0	0.09	1.00	+0.3
58	0.0	+1.0	0.00	1.00	0.0	03	−0.3	+1.0	0.09	1.00	−0.2
59	−0.5	+0.8	0.25	0.64	−0.4	04	−0.7	+0.7	0.49	0.49	−0.5
1860	−0.9	+0.4	0.81	0.16	−0.4	1905	−1.0	+0.2	1.00	0.04	−0.2
61	−1.0	−0.1	1.00	0.09	+0.1	06	−0.9	−0.4	0.81	0.16	+0.3
62	−0.8	−0.6	0.64	0.36	+0.5	07	−0.6	−0.8	0.36	0.64	+0.5
63	−0.3	−0.9	0.09	0.81	+0.3	08	−0.1	−1.0	0.01	1.00	+0.1
1864	+0.2	−1.0	0.04	1.00	−0.2	1909	0.5	−0.9	0.25	0.81	−0.4

With this table the computation of x and y is so easy a matter that I have for the most part computed all the series of observations I could find which extended through as long a time as a single spot-period. Each station is generally treated separately; but in a few instances I have combined the results of neighboring stations into a single mean. I forego any detailed description of the various methods by which the material, even when accurate, had to be treated in order to obtain the best annual means, presumably referred to a uniform standard. The more important sources are those already cited.

The following table shows the observed annual deviations formed from my own work. But in addition to these I have included observations, often fragmentary, made at British colonial stations, and published in the British meteorological reports:

TABLE II.

Mean Annual Departures of Temperature at Stations or in Regions.

	U. S. Fahr.	Habana C.	Kingston Fahr.	Argentine C.	Bombay Fahr.	Madras Fahr.	Calcutta Fahr.	Ceylon Fahr.	Manila C.	Australia Fahr.	Batavia C.	Apia C.	Mean C.
1871	+0.8	.....	.....	.....	+0.5	+0.3	0.0	.....	.....	.....	− 0.5	.....	+0.18
72	−0.1	.....	.....	−0.04	−0.7	−0.3	+0.7	.....	.....	.....	− 0.4	.....	−0.12
73	+0.3	.....	.....	+0.13	−0.7	−0.3	+0.7	.....	.....	.....	− 0.2	.....	−0.03
74	+0.2	−0.1	.....	−0.29	−1.3	−0.7	−0.2	.....	.....	.....	− 0.7	.....	−0.28
1875	−0.3	+0.1	.....	−0.01	−0.3	+0.1	−0.2	.....	.....	.....	− 0.2	.....	−0.04
76	+0.2	−0.3	.....	0.00	−0.3	+0.2	+1.0	.....	.....	.....	− 0.2	....	−0.01
77	+0.4	−0.1	.....	+0.60	+0.7	+0.6	+0.1	.....	.....	.....	+ 0.4	.....	+0.25
78	+0.3	+0.1	.....	−0.28	+0.5	+1.2	+1.0	.....	.....	.....	+ 1.2	.....	+0.31
79	+0.2	−0.2	.....	+0.13	−1.0	−0.2	+0.9	.....	.....	.....	− 0.2	.....	−0.08
1880	−0.1	+0.2	.....	0.00	−0.1	−0.1	−1.0	.....	.....	.....	− 0.6	.....	−0.10
81	+0.6	+0.1	+0.2	+0.22	+0.4	+0.3	−0.5	+0.2	.....	.....	+ 0.2	.....	+0.14
82	+0.5	+0.4	0.0	−0.05	−0.5	0.0	0.0	−0.2	.....	.....	− 0.4	.....	+0.02
83	+0.6	+0.4	+0.2	+0.24	−0.8	−0.3	−1.0	−0.6	−0.2	−0.4	− 0.2	.....	−0.07
84	+0.4	0.0	−0.3	+0.25	−0.8	−0.7	−1.3	−0.5	−0.6	−0.1	− 0.3	.....	−0.23
1885	0.0	−0.5	+0.6	−0.02	−0.4	−0.2	+0.2	−0.1	−0.1	+0.2	− 0.1	.....	−0.08
86	−1.0	−0.2	+0.7	+0.16	−0.1	−0.5		0.0	−0.2	+0.1	+ 0.2	.....	−0.08
87	−0.2	0.0	−0.4	+0.38	−0.9	−0.4		−0.5	−0.2	0.0	− 0.5	.....	−0.14
88	−0.5	+0.2	+0.6	+0.64	+0.5	−0.2		+0.2	+0.1	+1.2	+ 0.4	.....	+0.20
89	+0.2	+0.2	+0.9	−0.22	0.0	0.0		+0.3	+0.6	+0.7	+ 0.8	.....	+0.22
1890	+0.9	+0.1	−0.6	−0.29	−0.3	−0.3		−0.2	−0.2	+0.6	− 0.4	−0.21	−0.07
91	−0.2	0.0	+0.2	+0.08	−0.1			−0.2	0.0	−0.4	+ 0.5	−0.24	−0.04
92	−0.8	−0.4	−0.7	−0.71	+0.3			−0.1	+0.1	−0.5	+ 0.1	−0.32	−0.23
93	−0.3	+0.2	−0.9	−1.3	−0.7			−0.6	−0.2	0.0	− 0.5	−0.62	−0.35
94	−0.3	−0.2	−0.8	−0.2	0.0			−0.2	−0.2	−0.6	− 0.2	−0.24	−0.20
1895	−1.0	−0.1	−0.2	+0.6	0.0			−0.1	−0.1	−0.1	+ 0.1	−0.32	−0.09
96	+0.4	0.0	+0.4	+1.1	+1.1			+0.4	0.0	−0.5	+ 0.6	−0.31	+0.19
97	+0.1	+0.2	+0.3	−0.2	−0.1			+0.7	+0.6	+0.1	+ 1.1	+0.21	+0.22
98	−0.1	−0.2	−0.6	−0.4	+0.7			+0.3	0.0	+0.5	+ 0.2	+0.07	0.00
99	−0.4	+0.3	.....	.....	+0.4			0.0	−0.2	−0.4	0.0	+0.20	+0.02
1900	+0.5	+0.1	.....	.....	+0.4			+1.1	+0.5	−0.7	+ 0.6	+0.36	+0.27
01	−0.9	−0.4	.....	.....	+0.2			+1.7	+0.1	+0.3	.....	+0.51	+0.14
02	−0.2	+0.1	.....	.....	+0.5			.....	−0.1	.....	.....	+0.50	+0.12
03	−0.7	−0.1	.....	.....	.....			.....	.....	.....	.....	+0.55	0.00
04	−0.6	.....	.....	.....	.....			.....	.....	.....	.....	+0.13	−0.08
Σ_2	8.1	5.5	4.6	8.5	8.8	4.1	5.1	4.6	4.3	4.3	12.0	4.8	.....
Mean v^2	0.24	0.18	0.26	0.32	0.28	0.21	0.34	0.22	0.21	0.23	0.40	0.32	.....
w	3	4	3	3	3	4	2	4	4	3	1	3	.....

The following example of the computation of the fluctuation from the annual departures at Kingston, Jamaica, will make the process clear :

Coefficient of Sunspot Fluctuation at Kingston, Jamaica.

Years	Mean Tempera-ture	Deviation	$a\Delta\tau$	$b\Delta\tau$	Σa	Σb	Σa^2	Σb^2	Σab
1881	79.0	+0.2	—0.1	+0.2	—3.4	—0.9	2.78	3.23	—0.6
82	78.8	0.0	0.0	0.0	+3.1	+1.7	2.79	3.07	—0.6
83	79.0	+0.2	—0.2	0.0	—2.7	—2.4	2.95	2.98	+0.3
84	78.5	—0.3	+0.3	+0.2					
					—3.0	—1.6	8.52	9.28	—0.9
1885	79.4	+0.6	—0.2	—0.5					
86	79.5	+0.7	+0.1	—0.7					
87	78.4	—0.4	—0.2	+0.3					
88	79.4	+0.6	+0.5	—0.2			Normal Equations :		
89	79.7	+0.9	+0.9	+0.2					
						8.52 x — 0.9 y — 3.0 z = + 2.1			
1890	78.2	—0.6	—0.4	—0.4		— 0.9 x + 9.28 y — 1.6 z = — 2.0			
91	79.0	+0.2	0.0	+0.2		— 3.0 x — 1.6 y + 18.0 z = — 0.4			
92	78.1	—0.7	+0.2	—0.6					
93	77.9	—0.9	+0.7	—0.5					
94	78.0	—0.8	+0.8	—0.1					
1895	78.6	—0.2	+0.2	+0.1					
96	79.2	+0.4	—0.2	—0.4			$x = + 0.23$		
97	79.1	+0.3	0.0	—0.3		Result	$y = — 0.20$		
1898	78.2	—0.6	—0.3	+0.5			$z = + 0.02$		
Mean	78.8	$\Sigma = —0.4$	+2.1	—2.0					

In addition to the observations collected by myself for this work, I have made use of those of Köppen cited in the paper already referred to. This course was adopted because it did not seem necessary to repeat Köppen's work, even were the means of doing so available, which was not the case for the earlier observations. So far as I could infer from an examination of his work, and its comparison with published records, it is practically complete for all our present purposes. It is possible that there is a slight duplication of some of the observations in the work of Köppen and of myself, arising from the fact that his series and mine in a few cases overlap. But these cases are too few to be important, and only amount to assigning double their proper weight to the few duplicated records.

The results for x and y, with the numbers necessary for their final combination, are shown in Table III. The first column gives the place, or in a few cases the region, in which the observations were made. Down to Barbadoes the temperatures were those worked up by myself. The nine following are the regions within which the deviations were given and discussed by Köppen.

The value of x and y are in all cases expressed in degrees Centigrade, although the original deviations were often expressed in degrees of the Fahrenheit scale.

TABLE III.

Coefficients Expressing Observed Fluctuations of Temperature through the Sunspot Periods at Various Places or in Various Regions in the Form: $\Delta\tau = x \cos v + y \sin v.$

Place	Period of Obs.	x	y	W	Σ^2	w	$\cos \phi$
Calcutta	1868–85	+0.22.	−0.06	9	0.48	6	1.0
Ceylon	1881–01	+0.23	+0.02	10	0.32	10	1.0
Bombay	1840–01	+0.09	−0.01	28	0.37	25	1.0
Madras	1861–9	+0.14	+0.08	15	0.28	16	1.0
Manila	1883–02	+0.20	0.00	10	0.08	12	1.0
Scutari	1865–86	−0.14	−0.09	11	1.53	3	0.9
Malacca, etc.	1893–03	+0.17	−0.23	5	0.30	5	1.0
Apia	1890–04	+0.30	+0.07	7	......	15	1.0
Mauritius	1885–96	−0.12	+0.11	6	0.22 .	9	1.0
Natal	1872–86	+0.25	−0.08	7	1.90	1	0.9
Batavia	· 1866–00	+0.27	−0.08	17	0.23	20	1.0
Australia	1883–01	+0.17	−0.05	9	0.24	13	0.8
Malta	1865–81	+0.06	−0.02	9	0.26	11	0.8
Gibraltar	1854–82 .	−0.07	−0.31	12	1.20	3	0.8
Washington	1871–04	+0.14	+0.09 .	17	1.25	4	0.8
Key West	1871–04	−0.21	+0.01	17	0.71	8	0.9
St. Louis	1871–04	+0.28	+0.16	17	1.70	3	0.8
Galveston	1871–04	−0.06	+0.07	17	0.76	8	0.9
San Diego	1871–04	+0.17	−0.09	17	·1.01	5	0.9
Bermuda	1856–79	−0.28	−0.29	7	0.78	2	0.9
Havana	1874–03	0.00	+0.06	15	0.07	10	0.9
Kingston	1881–98	+0.13	−0.12	9	0.31	10	0.9
Barbadoes	1865–82	+0.14	−0.43	9	2.07	1	0.9
S. Africa	1842–67	+0.02	−0.08	12	0.13	20	0.8
Trop. America	1824–69	+0.16	−0.07	22	0.24	·25	1.0
S. U. States	1823–59	−0.04	+0.14	18	0.67	9	0.8
Farthest India	1820–62	+0.29	−0.08	21	0.17	25	1.0
India & Sunda	1840–58	+0.04	−0.20	9	0.14	·10	1.0
China—Japan	1841–55	+0.29	+0.53	8	0.26	· 10	0.7
Temp. S. Amer.	1843–60	− 0.06	−0.19	9	0.11	15	0.8
Australia	1841–70	+0.08	0.00	15	0.13	20	0.8
Mediterranean	1820–70	+0.11	−0.02	26	0.20	25	0.7

Column W gives approximately the integral part of the coefficient aa or bb. In the case of observations extending through any integral number of periods these two values would be the same. Practically they are always so nearly the same, approximately half the number of years, that it was unnecessary to make any distinction between them. In other words, the values of x and y may be regarded as always of equal weight.

Were the accidental fluctuations at the several stations equal in amount, W would be the weight to assign to each result. But, as a matter of course, different points and different regions are subject to different mean fluctuations. The mean of the squares of these fluctuations is shown in the column Σ^2. In a rigorous treatment by the method of least squares the value of Σ should be derived from the residuals left when the concluded values of the unknown quantities are substituted in the equation of condition. But, for obvious reasons, we should not find the residuals from each special solution, but by substituting the final values of the unknowns derived from the combination of all the data. Even then the weight might frequently be illusory, through a purely fortuitous accordance of the observations with the final results. Actually, therefore, I have deemed it best to use for Σ^2 simply the mean square of the actually observed deviations from the normal.

The weights to be assigned will then be proportional to $W \div \Sigma^2$. In order to express them in convenient units I have put approximately for the weight $w = W \div 3\Sigma^2$. This formula has not however been without some modifications as will be seen by the columns W, Σ^2, and w. Owing to the possibility of systematic errors at any one station the stations which by the formulae would be entitled to great weight have their weights slightly diminished, and no station is allowed a greater weight than 25.

It was found by Köppen that the fluctuation was greatest in the tropics, and diminished in either direction as the latitude increased. This is what we should expect. We may therefore plausibly suppose its amount at any place to be proportional to the insolation, or to the cosine of the latitude. The value of this cosine to a single place of decimals is given in the last line of the table.

It now remains from all the numbers of this table to derive the most probable values of x and y for the equatorial regions. The values given are those derived from observation in each region, without correction for latitude. Putting x_0 and y_0 for the values at the equator we form from each given x and y an equation of condition in the form

$$x_0 \cos \phi = x$$
$$y_0 \cos \phi = y$$

The final values are

$$x_0 = \frac{\Sigma wx \cos \phi}{\Sigma w \cos^2 \phi}$$

with a corresponding expression for y.

We find, from the numbers of the table,

$$\Sigma wx \cos \phi = + 37.5 \qquad \therefore x_0 = 0.\overset{\circ}{1}3$$
$$\Sigma wy \cos \phi = - 6.2 \qquad y_0 = -0.02$$
$$\Sigma w \cos^2 \phi = 298 \qquad \rho = 0.13$$
$$c = 9°$$

Hence, for the sun-spot fluctuation :

$$\Delta\tau = 0°.13 \cos \mu t - 0°.02 \sin \mu t = 0°.13 \cos (\mu t + 9°)$$

The expression has been derived without any reference to the actual epochs of the solar spottedness. All that we have done is to assume a period of 11.13 years in the temperature, and determine what constants of a harmonic fluctuation in this period will best represent the observations. It now remains to compare the epochs of temperature thus derived with those of the spots. This is done in Table IV. In studying this table it must be noted that the given epochs are not those derived individually from the observations in each case, but are the results of the general formulae which best represent all the observations. Consequently, the difference between the sunspot epochs and the temperature epochs as derived are constant in each of the respective phases of maxima and minima.

TABLE IV.

Comparison of Epochs of Temperature and Sun-spots.

Max. Temp.	Min. ⊙ Spots	Δ	Min. Temp.	Max. ⊙ Spots	Δ
1844.3	44.6	-0.3	1849.9	49.3	$+0.6$
1855.4	55.8	-0.4	1861.0	60.4	$+0.6$
1866.6	66.9	-0.3	1872.1	71.5	$+0.6$
1877.7	78.0	-0.3	1883.3	82.6	$+0.7$
1888.8	89.2	-0.4	1894.5	93.8	$+0.7$
1900.0	00.3	-0.3	1905.6	04.9	$+0.7$

It will be seen that, in the general mean of all the epochs, the temperature epoch follows the spot epoch, the comparisons of each phase being

Maximum temperature — minimum sunspots $- 0.33$

Minimum temperature — maximum sunspots $+ 0.65$

Mean of all the comparisons $+ 0.16$

The difference between the comparisons of the two phases arises from the fact that, by the method adopted, the intervals between the maxima and minima of temperatures necessarily come out equal, while those between the maxima and minima of sunspots are unequal.

The general conclusion is that the fluctuations of temperature follow very closely those of the sunspots, according to the law first clearly brought out by Köppen. The slight lagging of $0^y.16$, or two months, is too small to be regarded as the result of anything but accidental deviations, being less than the probable error of its amount.

Very remarkable is the fact that the actual fluctuation is less than half that found by Köppen. In order to show whether, when treated by the more rigorous method, the deviations of temperature used by him would give a different result from mine, we have only to find the general result of his data taken separately. This we do by deriving the mean values of x and y from his data alone, the individual results of which are found in the last nine lines of the table. These give

$$x = 0°.13 \qquad y = - 0°.05$$

Accidentally, the principal coefficient of the fluctuation is practically the same whether derived from his observations or from the others.

Although the reality of this 11-year fluctuation seems to be placed beyond serious doubt, the amplitude being several times its probable error, its amount is too small to produce any important direct effect upon meteorological phenomena.

§ 9. Study of Irregular Fluctuations of the Mean Annual Temperature.

The next question before us is whether, after correcting the annual departures of temperature for the sun-spot inequality, indications can be found of fluctuations in the general temperature other than those arising from accidental deviations. In this study we apply the statistical method developed in Chapter II, § 4, preceding. The data are shown in Table V, which is formed from Table II by reducing to the centi-

TABLE V.

Reduced Annual Deviations of Temperature at Stations or in Regions in Degrees C.

Year	U. S.	Habana	Kings-ton	Argen-tina	Bombay	Madras	Calcutta	Ceylon	Manila	Aus-tralia	Batavia	Apia	India
1871	+.52	°	°	°	+.42	+.32	+.12	°	°	°	−0.38	°	+.31
.72	+.03	....	....	+0.09	−.27	−.07	+.53	.	....	....	−0.27	....	.00
73	+.29	....	....	+0.22	−.31	−.11	+.49	....	....	....	−0.11	....	−.04
74	+.13	−.07	....	−0.26	−.67	−.37	−.07	....	....	....	−0.67	....	−.40
1875	−.24	+.06	....	−0.05	−.24	+.06	−.14	....	....	....	−0.24	....	−.08
76	.00	−.40	....	−0.10	−.30	.00	+.50	....	....	....	−0.30	....	+.01
77	+.07	−.23	....	+0.47	+.27	+.17	−.03	....	....	....	+0.27	....	+.16
78	+.08	−.02	....	−0.40	+.18	+.58	+.48	....	....	....	+1.08	....	+.42
79	+.03	−.27	....	+0.06	−.67	−.17	+.43	....	....	....	−0.27	....	−.20
1880	−.10	+.20	....	0.00	−.10	−.10	−.60	....	....	....	−0.60	....	−.21
81	+.37	+.17	+.17	+0.29	+.27	+.27	−.23	+.17	....	....	+0.27	....	+.16
82	+.42	+.52	+.12	+0.07	−.18	+.12	+.12	+.02	....	....	−0.28	....	+.02
83	+.43	+.53	+.23	+0.37	−.27	−.07	−.47	−.17	−.07	−.07	−0.07	....	−.21
84	+.30	+.10	−.10	+0.35	−.30	−.30	−.60	−.20	−.50	.00	−0.20	....	−.32
1885	+.04	−.46	+.34	+0.02	−.16	−.06	+.14	−.06	−.06	+.14	−0.06	....	−.05
86	−.63	−.23	+.37	+0.13	−.13	−.33	....	−.03	−.23	+.07	+0.17	....	−.17
87	−.19	−.09	−.29	+0.29	−.59	−.29	....	−.39	−.29	−.09	−0.59	....	−.41
88	−.43	+.17	+.17	+0.51	+.17	−.23	....	−.03	−.03	+.57	+0.27	....	−.05
89	−.02	+.08	+.38	−0.34	−.12	−.12	....	+.08	+.48	+.28	+0.68	....	−.05
1890	+.42	+.02	−.38	−0.37	−.28	−.28	....	−.18	−.28	+.22	−0.48	−.29	−.24
91	−.11	−.01	+.09	+0.07	−.11	....	....	−.11	−.01	−.21	+.049	−.25	−.11
92	−.34	−.34	−.34	−0.65	+.26	....	....	−.04	+.16	−.24	+0.16	−.26	+.09
93	−.09	+.31	−.39	−1.19	−.29	....	....	−.19	−.09	+.11	−0.39	−.51	−.23
94	−.07	−.07	−.27	−0.07	+.13	....	....	+.03	−.07	−.17	−0.07	−.11	+.07
1895	−.49	+.01	+.01	+0.71	+.11	....	....	+.01	+.01	+.01	+0.21	−.21	+.05
96	+.25	+.05	+.25	+1.15	+.65	....	....	+.25	+.05	−.25	+0.65	−.26	+.42
97	+.08	+.18	+.18	−0.22	−.12	....	....	+.38	+.58	+.08	+1.08	+.19	+.17
98	−.19	−.29	−.39	−0.49	+.31	....	....	+.11	−.09	+.21	+0.11	−.02	+.20
99	−.33	+.17	....	....	+.07	....	....	−.13	−.33	−.33	−0.13	+.07	−.04
1900	+.18	−.02	....	....	+.08	....	....	+.48	−.38	−.52	+0.48	+.24	+.31
01	−.59	−.49	....	....	+.01	....	....	+.81	+.01	+.11	....	+.42	+.47
02	−.12	+.08	....	....	+.28	....	....	....	−.12	....	....	+.48	+.28
03	−.35	−.05	....	....	....	....	....	....	....	....	....	+.60	....
1904	−.19	....	....	....	....	....	....	....	....	....	....	+.24	....
Wt. *w*	3	4	3	3	3	4	2	4	4	3	1	3	4

grade scale and correcting the departures for the sun-spot fluctuation. They are given in some cases for the individual stations, and in others for entire regions. The column "India" is the weighted mean of the four Indian stations alone, which has been separately formed for a reason which will hereafter be shown.

In combining the departures into a general mean it is advisable to assign different weights to different stations, on account of the diversity of the mean fluctuation, as shown in the several columns. If we could regard each departure as independent of all the others, and free from any source of systematic error, the weights would be proportional to the inverse square of the mean fluctuations, as given in each column. But this course would result in giving too great a relative weight to the stations of small fluctuation. Actually, in the first combination, the weights used are those at the bottom of the several columns.

TABLE VI.

Treatment of Annual Departures.

Year	$\tau_1{}'$	τ_1	$\Delta\odot$	τ'	W'	$W'^2\tau'^2$	Σw^2v^2	τ	W	$W^2\tau^2$	Σw^2v^2
1871	+0.18	+0.19	−0.12	+0.30	13	15.2	5.7	+0.31	8	5.76	4.16
72	−0.12	−0.12	−0.13	+0.01	16	0.0	2.0	+0.01	11	0.01	0.19
73	−0.03	+0.03	−0.09	+0.06	16	0.9	3.2	+0.12	11	1.69	1.35
74	−0.28	−0.23	−0.03	−0.25	20	25.0	7.4	−0.20	15	9.00	3.94
1875	−0.04	−0.04	+0.04	−0.08	20	2.6	1.2	−0.08	15	1.44	0.70
76	−0.01	−0.05	+0.10	−0.11	20	4.8	4.6	−0.15	15	4.84	2.74
77	+0.25	+0.24	+0.13	+0.12	20	5.8	3.8	+0.11	15	2.56	3.26
78	+0.31	+0.24	+0.12	+0.19	20	14.4	8.1	+0.12	15	2.89	5.59
79	−0.08	−0.05	+0.07	−0.15	20	9.0	6.2	−0.12	15	3.61	1.99
1880	−0.10	−0.06	0.00	−0.10	20	4.0	2.7	−0.06	15	0.81	1.73
81	+0.14	+0.16	−0.07	+0.21	27	32.1	4.8	+0.23	18	16.81	3.21
82	+0.02	+0.08	−0.12	+0.14	27	14.3	6.5	+0.20	18	14.44	6.40
83	−0.07	+0.02	−0.13	+0.06	34	4.2	9.8	+0.15	25	14.44	8.58
84	−0.23	−0.16	−0.10	−0.13	34	19.5	9.8	−0.06	25	1.96	8.00
1885	−0.08	−0.07	−0.04	−0.04	34	1.8	5.1	−0.03	25	0.49	4.51
86	−0.08	−0.07	+0.03	−0.11	32	12.4	7.1	−0.10	25	6.25	7.17
87	−0.14	−0.10	+0.09	−0.23	32	54.2	10.1	−0.19	25	22.09	6.59
88	+0.20	+0.24	+0.13	+0.07	32	5.0	7.9	+0.11	25	7.29	7.31
89	+0.22	+0.26	+0.12	+0.10	32	10.2	7.9	+0.14	25	12.25	7.09
1890	−0.07	−0.05	+0.08	−0.15	35	27.6	8.7	−0.13	28	12.96	7.88
91	−0.04	−0.04	+0.01	−0.05	31	2.4	1.6	−0.05	28	1.00	1.48
92	−0.23	−0.26	−0.06	−0.17	31	27.8	10.1	−0.20	28	31.36	9.26
93	−0.35	−0.35	−0.11	−0.24	31	55.4	20.0	−0.24	28	46.24	19.40
94	−0.20	−0.21	−0.13	−0.07	31	4.7	1.2	−0.08	28	5.76	1.34
1895	−0.09	−0.09	−0.11	+0.02	31	0.4	7.2	+0.02	28	0.16	7.10
96	+0.19	+0.17	−0.05	+0.24	31	55.4	19.2	+0.22	28	36.00	17.00
97	+0.22	+0.22	+0.02	+0.20	31	48.1	10.7	+0.20	28	31.36	8.66
98	0.00	−0.03	+0.09	−0.09	31	7.8	7.0	−0.12	28	12.25	6.67
99	+0.02	−0.01	+0.13	−0.11	25	7.6	4.6	−0.14	22	7.29	4.27
1900	+0.27	+0.25	+0.12	+0.15	25	14.1	9.4	+0.13	22	7.29	7.25
01	+0.14	+0.08	+0.09	+0.05	24	1.4	19.3	−0.01	21	0.09	12.63
02	+0.12	+0.13	+0.02	+0.10	17	2.9	3.3	+0.11	18	3.61	3.67
03	0.00	0.00	−0.05	+0.05	10	0.3	4.3	+0.05	10	0.25	4.49
1904	−0.08	−0.08	−0.11	+0.03	6	0.0	0.9	+0.03	6	0.01	0.85

The process of applying the criterion for correlation is shown in Table VI. To illustrate the method as fully as possible, two combinations of the data have been made. In the first the four Indian stations are treated as independent, in the second their mean is used as a single region. The second and third columns show the general mean departures of temperature, uncorrected for the sun spot fluctuations, as formed from the departures in Table II. In the first of these the four Indian stations are treated as if they were independent; in the second their combined mean is used, as found from the last column of Table V.

$\Delta \odot$ is the sun-spot fluctuation. Subtracting it from the two columns of means we have a world-departure of temperature, found in the columns τ' and τ, according to the use of the Indian stations. Following each of these is its weight, which is the sum of the weights of the individual departures.

Fixing our attention on these world-departures we note that their general mean value is about $\pm 0°.13$, and that in only 7 of the 34 years does it rise to $0°.2$. If we could regard these departures as actual means for the entire globe, they would indicate corresponding fluctuations in the sun's radiation. But, before we can draw any conclusion to this effect, we must determine whether the departures exceed in their general mean the values to be expected from the accidental deviations in the separate regions.

As the statistical method has been set forth, the sum of the squares of the general deviations τ are derived from any unbroken series of observations at a number n of stations extending through a number r of years. In substance, the method consists in subtracting from the sum of the squares of the products $W\tau$ the portions of the squares which would be due to the accidental deviations, or $\Sigma w^2 v^2$. The remainder $\Sigma W^2 \tau^2 - \Sigma w^2 v^2$, which we have called Δ, is proportional to the sum of the squares of the deviations for the whole globe, as shown by the equations (16) and (17). We might subtract for each unbroken series, not the squares of the actual regional deviations v, but the product of the mean values of v^2 by r. The final result would obviously be the same in either case.

We now sum the columns $W'^2 \tau'^2$ and $w'^2 v'^2$ to find the value of Δ, dividing the time into convenient terms of four or five years as follows:

Term	$W'^2\tau'^2$	$w'^2v'^2$	Δ	W'^2
1871–74	41.1	18.3	+ 22.8	1081
1875–79	36.6	23.9	+ 12.7	2000
1880–84	74.1	33.6	+ 40.5	4170
1885–89	83.6	38.1	+ 45.5	5252
1890–94	117.9	41.6	+ 76.3	5069
1895–99	119.3	48.7	+ 70.6	4469
1900–04	18.7	37.2	− 18.5	1626
Sum	491.3	241.4	249.9.	23667

The weight assigned to each station and region being taken as constant we have

$$\Sigma_{i,j} w^2 = r_1 w_1^2 + r_2 w_2^2 + \cdots + r_n w_n^2$$

r being, in each case, the number of years through which the observations extend.

To find the mean cosmical fluctuation indicated we have, for use in (17)

$$\Sigma W'^2 - \Sigma^2 w^2 = 20766$$

$$\text{Mean } \tau_0^2 = \frac{\Delta}{\Sigma W^2 - \Sigma^2 w^2} = .012$$

$$\therefore \text{mean } \tau_0 = \pm\, 0^\circ.11 \text{ C.}$$

This is the mean general fluctuation of temperature of the earth from year to year which is indicated by the data of observation.

But, before we accept this as really cosmical, we must find whether it affects all the stations, or whether the correlation exists only between stations so situated that they may be subject to like departures of temperature through the great movements of the air from one region to another.

The four Indian stations are especially in close proximity; we shall therefore discuss their departure by themselves, to decide whether they show any well-marked correlation. In doing this it will be unnecessary to make any distinction of weights. We shall therefore put $w = 1$ in each case, which will make W identical with the number of stations. Of course we must then use for τ the unweighted means, which are slightly different from those of Table V. Starting with 1871, we find these to be $\tau = + 0^\circ.29, + 0^\circ.06, + 0^\circ.02$, etc., instead of $+ 0^\circ.31, 0^\circ.00, - 0^\circ.04$, etc. For use in the equation (9) the values of $n\tau^2$ are .252, .011, .001, etc. These we sum by periods during which the number of stations remains unchanged. Then we sum the individual departures in the same way, and divide each annual sum by n. We have for 1871, $\Sigma v^2 = .42^2 + .32^2 + .12^2 = 0.293$. This gives, for 1871, $\Sigma v^2 \div n = .098$, in using which two decimals are amply sufficient. Carrying through this computation for each year and summing by periods, we find the following results:

Period	n	$n^2 \Sigma \tau^2$	Σv^2	Δ
1871–80	3	4.7	1.0	3.7
1881–85	4	3.1	1.2	1.9
1886–90	3	2.4	1.0	1.4
1891–01	2	2.4	1.9	0.5
		12.6		7.5

The positive correlation shown by Δ is so clearly marked as to leave no doubt, a result which accords with what we might anticipate from the geographical proximity of the stations.

We next investigate the result when the four Indian stations are combined into a single mean, which is found in the last column of Table V. The general world-departure then found is shown in column τ of Table VI, and the computation of the two series whose difference shows the correlation is shown in the last two columns of the table. Summing by terms as before, we have the following numbers:

Term	$W^2\tau^2$	w^2v^2	Δ	W^2
1871–74	16.4	9.6	+ 6.8	531
1875–79	15.3	14.3	+ 1.0	1125
1880–84	48.5	27.9	+ 20.6	2123
1885–89	48.4	32.7	+ 15.7	3125
1890–94	97.3	39.4	+ 58.0	3920
1895–99	87.1	43.7	+ 43.4	3620
1900–04	11.2	28.8	− 17.6	1385
Sum	324.2	196.3	127.9	15829

We thus have for the entire period of investigation

$$\Delta = 127.9$$

The value of the general fluctuation is thus reduced to

$$\tau_0 = \pm\, 0°.07$$

a quantity not greater than its probable error.

But we still cannot assume that all the regions are so distant from each other as to be unaffected through an entire year by any common terrestrial cause, especially the winds. Considering first the proximity of the stations, we notice that Havana and Kingston may be regarded as in the same region with each other, and with the United States. Moreover, the Southeastern Asiatic and Australian stations are so linked in a geographical series that we cannot regard each as necessarily independent of that next to it. On the other hand North America, South America, Apia and the Asiatic-Australian series form four sets which we cannot deem to be correlated except through the action of a cosmical cause, presumably fluctuations of the sun's radiation, which would affect all the stations, how widely soever separated. We therefore inquire whether the correlation we have found is or is not quite general for the earth by correlating the stations in pairs by the method shown in § 5. Beginning with the widely separated stations we correlate the three North American regions, United

States, Havana and Kingston with the distant ones, shown in the following table with the result:

Correlation Between North American and Distant Regions, Taken Two by Two.

	U. S.		Havana		Kingston	
	$\frac{1}{2}\Delta$	$ww'r$	$\frac{1}{2}\Delta$	$ww'r$	$\frac{1}{2}\Delta$	$ww'r$
Argentina	+2.6	243	+2.6	300	+11.8	162
Apia	−4.3	135	−2.8	168	+ 3.5	81
Manila	+1.4	240	+0.3	320	+ 5.4	192
India	−0.8	384	−6.0	464	+ 2.3	216
Batavia	−1.6	90	−0.4	108	+ 3.5	54
Australia	−2.1	171	−1.9	228	+ 1.2	144
Sum	−4.8	1263	−8.2	1588	+27.7	849

The correlation between Argentina and distant regions is as follows:

Argentina:	Apia	$\frac{1}{2}\Sigma\Delta =$	+ 3°.6	$ww'r =$	81
"	Manila	"	− 4.6	"	192
"	India	"	+ 4.2	"	324
"	Batavia	"	+ 2.0	"	81
"	Australia	"	− 2.6	"	144
	Sum		+ 2.6		822

We have finally the correlation between Apia and the Indo-Australian regions.

Apia:	Manila	$\frac{1}{2}\Sigma\Delta =$	+ 2°.6	$ww'r =$	156
"	India	"	+ 5.8	"	156
"	Batavia	"	+ 1.0	"	33
"	Australia	"	− 0.3	"	108
	Sum		+ 9.1		453

The curious synchronism between the annual departures of Kingston and all the most distant stations, especially Argentina, may well excite notice. But I do not conceive that we can attribute it to anything but chance coincidence.

We next take the pairs between which we should expect correlation on account of their proximity. A detailed exhibit of the results does not seem necessary. The summation of $ww'vv'$ gives:

United States: Havana–Kingston	3 pairs	$\Delta = $ + 15°.7
Indo-Australian series	6 pairs	$\Delta = $ + 20.3

The complete summation of the values of $\bar{\Delta}$ gives 62°.3, in fair agreement with that derived from the combination of the squares of the deviations.

It seems therefore that, of the 36 pairs of regions, 9 which were in proximity contribute more than half to the making up Δ, the correlation number. Of these 9, the two extremes, India-Australia and Manila-Australia, are so distant from each other that they should be included in the class not subject to any common cause of change of temperature. Their contributions to $\frac{1}{2}\Delta$ are:

India-Australia	$\frac{1}{2}\Delta = -3.0$	$ww'r = 228$
Manila-Australia	" $= -0.7$	" $= 228$
Sum	-3.7	456

We now have the following equations for the mean value of that portion of the fluctuations of mean annual temperature which we may attribute to a general cause affecting the whole earth:

United States and dist. points			6 pairs	$1263\tau_0^2 =$	-4.8
Havana	"	"	"	6 "	1588	-8.3
Kingston	"	"	"	6 "	849	$+27.7$
Argentina	"	"	"	5 "	822	$+2.6$
Apia	"	"	"	4 "	453	$+9.1$
Australia	"	"	"	2 "	456	-3.7
Total			29	$5431\tau_0^2$	$+22.6$

This gives

$$\tau_0^2 = .0042$$

$$\tau_0 = \pm 0.065 \text{ C}.$$

This fluctuation, if regarded as real, is too minute to produce any important meteorological effect. That it may well arise from the accidental deviations is shown by the fact that, had Kingston been omitted, τ_0^2 would have come out negative, indicating a tendency toward an equalization of the general temperature of the globe from year to year. But there is nothing to justify us in rejecting Kingston for this reason, though a careful analysis might show that we have given it greater relative weight than it is entitled to. The same remark would, however, apply to Havana, the result of which is markedly in the opposite direction from that of Kingston.

§ 10. *Time Correlations in Annual World Temperatures.*

Returning to Table VI, it is noticeable that the larger outstanding departures τ do not seem to be scattered at random, but are rather collected in groups of like algebraic sign, as if they were the result of a fluctuation having a period of several years. It would be easy to represent them by the ordinates of a sinuous curve, but a conclusion based on this method would be altogether unreliable. We shall therefore apply the method of time correlation, developed in § 6, which will bring out with numerical exactness any period that may exist, or any periodic tendency. The numerical process is shown in the following lines, the numbers of which are formed as follows: Starting with the departure for 1871, 0.31, which, for our present purpose we call a_0, we form its square, and also its product by the following departures in the order of time to any extent to which we may suspect a correlation. In the present case we have considered it sufficient to form the products through terms of nine years. The nine consecutive products formed by multiplying 0.31 by τ for the years 1871 to 1880 are written in the first line of the following table.

Next we take the year 1872, form the square of its departure, and the products by the departure for the nine years following. These form the second line of the table. We repeat the process for 1873 and subsequent years to the end of the series and write the results in consecutive lines with each initial year of the series. Of course the number of years available will fall off by one for each line in which the initial year is greater than 1895. The series terminating with 1904, we have eight products for 1896, seven for 1887, etc.

TABLE VII.

Time Correlation Among Annual Temperatures.

Initial year	a_0^2	a_0a_1	a_0a_2	a_0a_3	a_0a_4	a_0a_5	a_0a_6	a_0a_7	a_0a_8	a_0a_9
1871	.096	+.003	+.037	−.062	−.025	−.046	+.034	+.037	−.037	−.019
72	.000	+.001	−.002	−.001	−.001	+.001	+.001	−.001	−.001	+.002
73	.014	−.024	−.010	−.018	+.013	+.014	−.014	−.007	+.028	+.024
74	.040	+.016	+.030	−.022	−.024	+.024	+.012	−.046	−.040	−.030
1875	.006	+.012	−.009	−.010	+.010	+.005	−.018	−.016	−.012	+.005
76	.022	−.016	−.018	+.018	+.009	−.035	−.030	−.022	+.009	+.004
77	.012	+.013	−.013	−.007	+.025	+.022	+.016	−.007	−.003	−.011
78	.014	−.014	−.007	+.028	+.024	+.018	−.007	−.004	−.012	.023
79	.014	+.007	−.028	−.024	−.018	+.007	+.004	+.012	+.023	−.013
1880	.004	−.014	−.012	−.009	+.004	+.002	+.006	+.011	−.007	−.008
81	.053	+.046	+.034	−.014	−.007	−.023	−.044	+.025	+.032	−.030
82	.040	+.030	−.012	−.006	−.020	−.038	+.022	+.028	−.026	−.010
83	.023	−.009	−.005	−.015	−.028	+.017	+.021	−.020	−.008	−.030
84	.004	+.002	+.006	+.011	−.007	−.008	+.008	+.003	+.012	+.014
1885	.001	+.003	+.006	−.003	−.004	+.004	+0.01	+.006	+.007	+.002
86	.010	+.019	−.011	−.014	+.013	+.005	+.020	+.024	+.008	−.002
87	.036	−.021	−.027	+.025	+.010	+.038	+.046	+.015	−.004	−.042
88	.012	+.015	−.014	−.006	−.022	−.026	−.009	+.002	+.024	+.022
89	.020	−.018	−.007	−.028	−.034	−.011	+.003	+.031	+.028	−.017
1890	.017	+.007	+.026	+.031	+.010	−.003	−.029	−.026	+.016	+.018
91	.003	+.010	+.012	+.004	−.001	−.011	−.010	+.006	+.007	−.007
92	.040	+.048	+.016	−.004	−.044	−.040	+.024	+.028	−.026	+.002
93	.058	+.019	−.005	−.053	−.048	+.029	+.034	−.031	+.002	−.026
94	.006	−.002	−.018	−.016	+.010	+.011	−.010	+.001	−.009	−.004
1895	.000	+.004	+.004	−.002	−.003	+.003	.000	+.002	+.001	+.001
96	.048	+.044	−.026	−.031	+.029	−.002	+.024	+.011	+.007	....
97	.040	−.024	−.028	+.026	−.002	+.022	+.010	+.006	....	....
98	.014	+.017	−.016	+.001	−.013	−.006	−.004	....	....	....
99	.020	−.018	+.001	−.015	−.007	−.004	....	....	....	....
1900	.017	−.001	+.014	+.007	+.004	....	....	....	....	....
01	.000	−.001	−.001	.000	....	....	....	....	....	....
02	.012	+.006	+.003	....	....	....	....	....	....	....
03	.003	+.002	....	....	....	....	....	....	....	....
04	.001	....	....	....	....	....	....	....	....	....
Σ	.700	+.162	−.080	−.209	−.147	−.031	+.111	+.068	+.019	−.278

It is to the summation found at the bottom of this table that our attention will be especially directed. It must be admitted that the periodicity among the numbers seems to be very well marked, the apparent period being about six years. This is so nearly one-half that of the sun-spot period that, if the result is not purely fortuitous, we may well regard this as an actual period.

Assuming the correlation to be real, the fact brought out may be found by dividing the first sum $[a_0^2]$ into each of the sums following. This is done in Table IX. The second column of the table gives the values of $[a_0a_0]$, $[a_0a_1]$, $\cdots$, $[a_0a_9]$, which are the sums Σ just found. The third column gives the quotients $[a_0a_i] \div [a_0a_0]$. Accepting them as real, the result may be expressed as follows: Whatever the mean annual world departure in any one year, we have had since 1871, as a mean rule, a departure in the same direction of 0.23 of its amount the year following. In the third year following we have had a departure in the opposite direction of 0.30, of its initial amount; in the fourth year of 0.21; in the sixth year a departure, now in the original direction, of 0.16; and in the ninth a departure in the opposite direction, of 0.40 of the initial departure.

To estimate the probability that this periodicity is real we must estimate the probable accumulated amount of the purely fortuitous deviations. We have for this purpose

Standard annual deviation $= \pm 0.14$

The probable mean value of a product of two such deviations will depend upon the law of statistical distribution. Our best result will be derived not by assuming the normal law of distribution, which may not be strictly applicable, but by taking the indiscriminate average, without regard to sign, of the entire 261 products. We thus find

General average aa $= .0155$

The average expected accumulations of 30 such sums, if fortuitous, will be about $\pm .08$. This, then, is the expected average value of a non-systematic $[a_0a_i]$ ($i = 1, 2, 3,$ etc.), for the period 1871–1904. The actual average we see to be 0.13. The excess is no greater than might well be the result of chance deviations. But the inference of its reality is strengthened by the evident 6-year periodicity of the sums. On the other hand, the existence of this period as an unbroken one is negatived by the fact that during the last ten years of the series the epoch is practically reversed. The proof of a permanent period half that of the sun-spots therefore falls to the ground. If there is any real periodicity the case is similar to that of the waves of the ocean when, after a series of definite period, a new series sets in with the same period, but not a continuation of the first.

TABLE VIII.

Time Correlations Through Nine-year Terms.

	Years 1871–1904			Years 1820–69	
i	$[a_0 a_i]$	Quot.	i	$[a_0 a_i]$	Quot.
0	+0.700	+1.00	0	+4.99	+1.00
1	+0.162	+0.23	1	+1.73	+0.34
2	−0.080	−0.11	2	+1.74	+0.35
3	−0.209	−0.30	3	+1.09	+0.22
4	−0.147	−0.21	4	+0.29	+0.06
5	−0.031	−0.04	5	+0.42	+0.08
6	+0.111	+0.16	6	+1.52	+0.30
7	+0.068	+0.10	7	−0.08	−0.02
8	+0.019	+0.03	8	−0.15	−0.03
9	−0.278	−0.40	9	−0.63	−0.13

The reality of the periodicity can be established only by carrying the investigation back through the years preceding 1871. I have done this with Köpping's table of annual departures already cited, after correction for the sun-spot inequality. The result is found in the second part of Table IX, preceding. There is here not only no periodicity, but, on the contrary, a tendency toward a persistence of the departure in the same direction for as much as six years. The products are, in general, several times larger than those for the modern period, showing wider accidental deviations. We may attribute both this and the systematic character of the correlation products to the imperfections of the older instruments and observations. But this would not be likely to mask entirely a six-year periodicity, if any such existed. We must, therefore, regard the seeming period as unreal, or at least open to serious doubt, notwithstanding the plausibility of the statistical evidence in its favor.

CHAPTER IV.

DISCUSSION OF MONTHLY DEPARTURES.

Since the only period exceeding a month that we can assign *a priori* as probable, that of the sun-spots, has already been investigated in the preceding chapter, the purpose of the present chapter is to determine whether the monthly departures of world-temperature show any systematic character not found in the results of the annual departures. If this result were the only one aimed at, ideal simplicity and perfection would require that we first correct the normal temperatures from which the departures are computed for the fluctuations already derived from the annual means. In other words, our normal temperature should include at least the sun-spot fluctuation. But this has not been done. Consequently, the general departures τ_0, affecting all parts of the world simultaneously, may be expected to reappear in the discussion and comparison of the monthly means. But it does not seem objectionable to allow this. We have only to recall the fact in drawing conclusions from any systematic departures that may be found.

The monthly mean departures which have been selected for discussion are partly those of Dove, and partly those specially collected for the present work. Among the latter are included those subsequently given in connection with the ten-day means.

§ 11. *Discussion of Dove's Departures.*

In the *Memoirs* of the Berlin Academy for 1858 Dove gives a great number of tables of observed temperatures at widely separated stations, which are in some points similar in form to those required for the present work. Those best adapted to the present purpose have therefore been used for material. These are found on pp. 364, etc., of the *Memoirs*. A certain number of regions were selected from Dove's tables so far apart that there seemed to be no possibility of a correlation of their monthly temperatures, except from some cosmical cause. It was also necessary to prefer stations and regions where the temperature was least subject to rapid fluctuations, and for reasons already mentioned, regions of low rather than of high latitude. The regions thus selected were :

Eastern Asia; mean of Nagasaki and Pekin.

Southern Europe; mean of stations in southern Russia.

United States; mean of several stations in the southern portion.

Cape of Good Hope; one station only.

Hobartown; one station only.

Madras; one station only.

In taking the means no distinction of weight was made between the different regions or stations.

The mean deviations formed from Dove's tables were tabulated and summed separately for each year. The observations at Hobartown terminated with September, 1848.

The results of the summation of the squares of the deviations for the several years are shown in the following table.

Dove's deviations are given in the Reaumer scale. For convenience these are used without change in the table.

TABLE IX.

Dove's Simultaneous Monthly Departures of Temperature from the Normal.

	E. Asia	S. Europe	U. S.	Cape	Hobarton	Madras	Σ	τ	τ^2
1845									
Jan.	+1.2	+1.8	+0.9	−1.1	+0.4	0.0	+3.2	+0.53	+0.28
Feb.	+0.3	−2.5	−0.1	−0.9	−0.4	−0.6	−4.2	−0.70	+0.49
Mar.	+0.5	−0.8	−0.1	+0.2	+0.1	−0.8	−0.9	−0.15	+0.02
April	−0.1	0.0	+1.3	−0.3	+0.4	−0.3	+1.0	+0.17	+0.03
May	+0.5	−1.6	−0.4	−0.1	−0.4	−0.1	−2.1	−0.35	+0.12
June	+0.7	+0.4	+0.2	−0.8	−0.1	+0.6	+1.0	+0.17	+0.03
July	0.0	+0.1	+0.5	−0.1	+0.8	+0.3	+1.6	+0.27	+0.07
Aug.	−0.6	−0.8	−0.2	−1.8	+0.2	−0.2	−3.4	−0.57	+0.32
Sept.	−1.0	0.0	0.0	−0.3	+0.9	+0.3	−0.1	−0.02	0.00
Oct.	−0.5	0.0	−0.4	+0.3	+0.7	−0.7	−0.6	−0.10	+0.01
Nov.	−0.6	+1.1	−0.9	−0.2	+0.2	−0.3	−0.7	−0.12	+0.01
Dec.	−2.4	+1.0	−2.3	−0.7	+0.2	−0.6	−4.8	−0.80	+0.64
1846									
Jan.	−0.6	+1.2	−0.3	−0.7	−0.2	....	−0.6	−0.12	+0.01
Feb.	−0.2	+1.3	−1.0	−0.2	−1.1	....	−1.2	−0.24	+0.06
Mar.	−0.8	+1.9	+0.1	−0.1	−0.5	....	+0.6	+0.12	+0.01
April	−0.2	+1.2	−0.6	−0.7	+0.1	....	−0.2	−0.04	0.00
May	−1.9	+0.9	+0.2	−0.6	+0.5	....	−0.9	−0.18	+0.03
June	−0.2	+1.5	−0.5	+0.1	+0.1	....	+1.0	+0.20	+0.04
July	−0.3	+1.2	−0.7	+1.5	−0.4	....	+1.3	+0.26	+0.07
Aug.	+0.5	+0.7	0.0	−0.7	+0.8	....	+1.3	+0.26	+0.07
Sept.	+0.9	+0.6	+0.1	+0.8	−0.2	....	+2.2	+0.44	+0.19
Oct.	+0.7	+0.9	−0.4	+1.1	+0.3	....	+2.6	+0.52	+0.27
Nov.	−0.3	−0.2	+0.2	+0.2	+0.5	....	+0.4	+0.08	+0.01
Dec.	+1.1	−1.1	+1.5	+1.1	+0.7	....	+3.3	+0.66	+0.44
1847									
Jan.	+0.8	−0.1	+0.6	+0.1	−0.2	....	+1.2	+0.24	+0.06
Feb.	−0.3·	−0.9	−0.5	+1.2	−0.2	....	−0.7	−0.14	+0.02
Mar.	−0.4	−0.6	−1.1	0.0	−0.4	....	−2.5	−0.50	+0.25
April	+1.1	−0.7	+0.5	−0.4	0.0	....	+0.5	+0.10	+0.01
May	−0.2	+1.8	−0.8	0.0	−0.8	....	0.0	0.00	0.00
June	−1.3	−1.8	−0.2	−0.1	−1.0	....	−4.4	−0.88	+0.77
July	+0.3	−0.2	−0.5	−0.7	+0.1	....	−1.0	−0.20	+0.04
Aug.	+1.0	0.0	−0.2	−0.8	+0.8	....	+0.8	+0.16	+0.02
Sept.	−0.9	−0.9	−0.3	0.0	+0.6	....	−1.5	−0.30	+0.09
Oct.	0.0	−0.8	+0.5	+1.0	−0.3	....	+0.4	+0.08	+0.06
Nov.	+2.0	−0.7	+1.0	−0.6	−1.2	....	+0.5	+0.10	+0.01
Dec.	+0.6	−0.4	−1.6	−1.0	+0.5	....	−1.9	−0.38	+0.14
1848									
Jan.	+0.2	−3.4	+0.9	+0.8	−0.9	....	−2.4	−0.48	+0.23
Feb.	−0.8	+0.3	+0.3	−0.5	−0.9	....	−1.6	−0.32	+0.10
Mar.	+0.6	+1.4	+0.1	+0.8	+0.3	....	+3.2	+0.64	+0.41
April	+0.7	+1.4	−0.5	−0.8	+1.6	....	+2.4	+0.48	+0.23
May	+0.8	−0.3	+0.3	+0.4	−0.1	....	+1.1	+0.22	+0.05
June	+0.7	+1.4	0.0	0.0	+0.2	....	+2.3	+0.46	+0.21
July	+0.8	−0.4	0.0	+0.2	−0.4	....	+0.2	+0.04	0.00
Aug.	0.0	+0.3	−0.1	−0.4	−0.3	....	−0.5	−0.10	+0.01
Sept.	+0.1	+0.1	−0.7	0.0	−0.7	....	−1.2	−0.24	+0.06
Oct.	+0.7	+0.4	+0.5	+0.4	....	....	+2.0	+0.50	+0.25
Nov.	−0.3	−0.4	−1.8	+0.4	....	....	−2.1	−0.52	+0.27
Dec.	+1.7	−0.4	+3.5	−0.1	....	....	+4.7	+1.17	+1.37
1849									
Jan.	+0.3	−0.3	+1.4	+0.2	....	....	+1.6	+0.40	+0.16
Feb.	+2.1	+0.7	−1.1	+0.4	....	....	+2.1	+0.52	+0.27
Mar.	+1.7	−0.3	+2.0	+0.4	....	....	+3.8	+0.95	+0.90
April	+0.6	−1.0	−0.2	0.0	....	....	−0.6	−0.15	+0.02
May	−0.6	0.0	+0.3	−0.4	....	....	−0.7	−0.17	+0.03
June	−0.5	+1.6	+0.2	+0.2	....	....	+1.5	+0.37	+0.14
July	−0.5	−0.3	−0.7	+0.2	....	....	−1.3	−0.32	+0.10
Aug.	+0.7	−0.8	+0.8	−0.1	....	....	+0.6	+0.15	+0.02
Sept.	+0.7	+0.4	+0.4	+0.1	....	....	+1.6	+0.40	+0.16

TABLE IX.—*Continued.*

Dove's Simultaneous Monthly Departures of Temperatures from the Normal.

	E. Asia	S. Europe	U. S.	Cape	Hobarton	Madras	Σ	τ	τ^2
1849									
Oct.	0.0	+0.4	+0.5	−0.2	....	....	+0.7	+0.17	+0.03
Nov.	−0.3	−0.9	+0.9	−0.5	....	....	−0.8	−0.20	+0.04
Dec.	0.0	−1.6	+2.9	−0.2	....	....	+1.1	+0.27	+0.07
1850									
Jan.	−0.4	−2.4	+2.7	−1.0	....	....	−1.1	−0.27	+0.07
Feb.	0.0	+0.8	−0.4	+0.6	....	....	+1.0	+0.25	+0.06
Mar.	+0.2	−1.5	+0.8	+0.2	....	....	−0.3	−0.08	+0.01
April	+1.4	+0.2	−0.1	+0.2	....	....	+1.7	+0.42	+0.18
May	−1.9	−0.6	−0.5	−0.8	....	....	−3.8	−0.95	+0.90
June	+0.6	−0.2	−0.9	−0.3	....	....	−0.8	−0.20	+0.04
July	−0.7	−0.7	+0.4	+0.1	....	....	−0.9	−0.22	+0.05
Aug.	−0.6	+0.2	+0.9	−0.1	....	....	+0.4	+0.10	+0.01
Sept.	−0.4	−0.8	+0.1	+0.3	....	....	−0.8	−0.20	+0.04
Oct.	+0.2	−1.7	−0.6	+0.2	....	....	−1.9	−0.47	+0.22
Nov.	−1.3	+0.8	+0.3	−1.0	....	....	−1.2	−0.30	+0.09
Dec.	−0.4	0.0	+1.1	−0.2	....	....	+0.5	+0.12	+0.01
1851									
Jan.	0.0	+0.9	+0.8	+0.5	....	....	+2.2	+0.55	+0.30
Feb.	0.0	−0.2	+1.6	+0.2	....	....	+1.6	+0.40	+0.16
Mar.	+0.2	−0.2	0.0	+0.5	....	....	+0.5	+0.12	+0.01
April	−1.5	+0.9	0.0	−0.1	....	....	−0.7	−0.17	+0.03
May	−0.7	−2.0	+0.1	+0.1	....	....	−2.5	−0.62	+0.38
June	−1.1	−0.5	+0.3	+0.4	....	....	−1.5	−0.37	+0.14
July	+0.3	−1.1	0.0	−0.3	....	....	−1.1	−0.27	+0.07
Aug.	−0.3	−0.7	−0.1	−1.3	....	....	−2.4	−0.60	+0.36
Sept.	−1.0	−1.5	−0.6	+0.1	....	....	−3.0	−0.75	+0.56
Oct.	−0.6	+0.8	0.0	+0.2	....	....	+0.4	+0.10	+0.01
Nov.	+0.2	−2.0	−0.2	0.0	....	....	−2.0	−0.50	+0.25
Dec.	−0.9	−1.1	−0.9	+0.3	....	....	−2.6	−0.65	+0.42
1852									
Jan.	−0.5	+1.2	−3.8	−0.3	....	0.0	−3.4	−0.68	+0.46
Feb.	−1.2	+0.6	+0.4	−0.7	....	+0.2	−0.7	−0.14	+0.02
Mar.	−2.3	−1.0	+0.9	−0.3	....	+0.4	−2.3	−0.46	+0.21
April	+0.2	−1.5	−0.9	+0.2	....	0.0	−2.0	−0.40	+0.16
May	−0.2	0.0	+0.4	+0.1	....	0.0	+0.3	+0.06	0.00
June	+0.3	−0.1	+0.3	−0.4	....	+0.1	+0.2	+0.04	0.00
July	+0.8	+0.3	−0.2	−0.1	....	−0.2	+0.6	+0.12	+0.01
Aug.	−0.2	+0.1	−0.1	+0.2	....	+0.1	+0.1	+0.02	0.00
Sept.	−0.4	+0.6	−0.2	+0.3	....	+0.4	+0.7	+0.14	+0.02
Oct.	+1.3	−0.2	+0.3	+0.1	....	+0.1	+1.6	+0.32	+0.10
Nov.	−1.0	+2.9	0.0	−0.1	....	+0.4	+2.2	+0.44	+0.19
Dec.	+0.4	+2.5	+1.2	0.0	....	+0.3	+4.4	+0.88	+0.77
1853									
Jan.	−0.2	+1.9	−1.1	+0.3	....	+0.9	+1.8	+0.36	+0.13
Feb.	−0.9	−0.6	0.0	+0.2	....	−0.5	−1.8	−0.36	+0.13
Mar.	−1.7	−1.6	+0.3	−0.4	....	+0.2	−3.2	−0.64	+0.41
April	−0.4	−1.8	+0.4	+0.2	....	−0.2	−1.8	−0.36	+0.13
May	0.0	−0.4	0.0	−0.2	....	+0.7	+0.1	+0.02	0.00
June	+0.1	−0.9	−0.4	+0.1	....	+1.2	+0.1	+0.02	0.00
July	+0.7	+0.6	+0.2	+0.4	....	+0.5	+2.4	+0.48	+0.23
Aug.	+0.7	+0.2	+0.4	0.0	....	+0.5	+1.8	+0.36	+0.13
Sept.	+1.2	+0.2	−0.1	−0.2	....	+0.9	+2.0	+0.40	+0.16
Oct.	−1.3	+0.5	+0.8	−0.7	....	+0.9	+0.2	+0.04	0.00
Nov.	0.0	+0.2	+1.3	+1.5	....	0.0	+3.0	+0.60	+0.36
Dec.	−0.6	−1.6	−1.5	+0.8	....	−0.2	−3.1	+0.62	+0.38
1854									
Jan.	−0.8	+0.9	+0.9	+1.1	....	−0.6	+1.5	+0.30	+0.09
Feb.	−1.6	−1.1	0.0	0.0	....	+0.4	−2.3	−0.46	+0.21
Mar.	+0.6	−0.2	+1.4	−0.4	....	+0.4	+1.8	+0.36	+0.13

TABLE IX.—*Concluded.*

Dove's Simultaneous Monthly Departures of Temperature from the Normal.

	E. Asia	S. Europe	U. S.	Cape	Hobarton	Madras	Σ	τ	τ^2
1854									
April	+0.6	−0.2	−1.5	+0.2	....	+1.0	+0.1	+0.02	0.00
May	+0.2	+0.4·	+0.2	+1.1	....	+0.8	+2.7	+0.54	+0.29
June	+0.3	−0.8	+0.2	+0.9	....	+1.3	+1.9	+0.38	+0.14
July	−0.4	−0.6	+0.4	−0.3	....	+0.6	−0.3	−0.06	0.00
Aug.	−0.3	−0.9	−0.1	+0.1	....	+0.7	−0.5	−0.10	+0.01
Sept.	0.0	−0.9	+0.7	+0.3	....	+0.4	+0.5	+0.10	+0.01
Oct.	+1.0	+0.1	+0.2	−0.2	.......·.	+0.4	+1.5	+0.30	+0.09
Nov.	+0.9	−1.2	−1.3	+1.1	....	+0.6	+0.1	+0.02	0.00
Dec.	+2.6	+2.2	−0.8	+0.4	....	+0.3	+4.7	+0.94	+0.88
1855									
Jan.	−1.8	−1.3	−0.1	+0.7	....	+0.2	−2.3	−0.46	+0.21
Feb.	−1.5	−0.6	−2.5	+0.3	....	+0.1	−4.2	−0.84	+0.70
Mar.	0.0	+0.6	−1.1	+0.7	....	−2.6	−2.4	−0.48	+0.23
April	+0.4	−0.9	+1.3	−0.1	....	+0.2	+0.9	+0.18	+0.03
May	−0.2	−0.8	−0.2	+0.6	....	+1.3	+0.7	+0.14	+0.02
June	+0.3	−0.6	−0.2	+0.3	....	+0.8	+0.6	+0.12	+0.01
July	0.0	0.0	+0.1	−0.4	....	+1.1	+0.8	+0.20	+0.04
Aug.	−1.7	+0.2	+0.3	0.0	....	+1.2	0.0	0.00	0.00
Sept.	−0.2	+0.3	+0.8	+0.1	....	+0.8	+1.8	+0.36	+0.13
Oct.	....	+2.0	−1.4	+0.6	....	+0.2	+1.4	+0.35	+0.12
Nov.	....	+0.2	+1.7	+1.7	....	+0.2	+3.8	+0.95	+0.90
Dec.	....	−2.7	+0.4	+1.2	....	....	−1.1	−0.37	+0.14
	....	....	....	....	....	....	....	....	+1.37

The sums of the squares of the deviations which enter into the theory are formed for each year, and shown in the following table. Σv^2 is, in each case, formed from the deviations in the preceding table. $\Sigma_r \tau^2$ is the sum from the last column of that table, which is multiplied, for each year, by n, the number of stations used. As shown in the general theory, the difference, $n^2 \Sigma \tau^2 - \Sigma_i v^2$, so far as it is not the result of accidental errors and deviations, measures the correlation among the stations.

Results of Dove's Mean Monthly Deviations.

Year	East Asia Σv_1^2	South Europe Σv_2^2	U. S. Σv_3^2	Cape Σv_4^2	Hobartown Σv_5^2	Madras Σv_6^2	Mean $\Sigma \tau^2$	$n^2 \Sigma \tau^2$	Σv^2	$n-1$	r	Equation for τ_0^2	Normal equation
1845	10.3·	15.7	9.3	6.7	2.7	2.6	2.13	77	47	5	12	$60\,\tau_0^2 = +\,4.9$	$360\,\tau_0^2 = +30$
46	7.7	15.6	4.7	7.2	3.4	....	1.23	31	39	4	12	$48\ -\ 1.5$	$240\ -\ 8$
47	10.1	10.3	6.9	5.1	4.7	....	1.47	37	37	4	12	$48\ -\ 0.1$	$240\ \ 0$
48	6.8	18.4	17.5	2.9	5.1	....	1.35	34	51	4	9	$36\ +\ 0.7$	$180\ -17$
49	9.7	8.7	18.1	0.9	....	....	2.11	34	37	3	12	$36\ -\ 0.9$	$144\ -\ 3$
1850	9.0	13.8·	11.8	3.4	....	....	1.78	28	38	3	12	$36\ -\ 2.4$	$144\ -10$
51	6.4	15.8	4.5	2.6	....	....	2.80	45	28	3	12	$36\ +\ 4.0$	$144\ +17$
52	10.8	20.2	18.1	1.0	....	0.7	2.00	50	51	4	12	$48\ -\ 0.2$	$240\ -\ 1$
53	8.4	14.0	6.4	4.0	....	5.2	2.13	53	38	4	12	$48\ +\ 3.1$	$240\ +15$
54	12.9	11.2	8.1	5.0	....	5.6	1.84	· 46	43	4	12	$48\ +\ 0.6$	$240\ +\ 3$
1855	8.7	15.7	15.2	6.4	....	12.6	2.72	68	58	4	12	$48\ +\ 1.9$	$240\ +10$
Σ	......	....	....	....	....	....	......	503	467	42	129	$492\,\tau_0^2 = +10.1$	$2412\,\tau_0^2 = +36$

By reduction to the centigrade scale the final equation becomes

$$2412\tau_0^2 = 56$$

This equation will be combined with those to be derived from the later material.   When taken alone it gives the result

$$\tau_0{}^2 = .023 ; \qquad \tau_0 = \pm\, 0°.15 \text{ C.}$$

§ 12. *General Discussion of Monthly Departures from 1872 to 1900.*

In pursuance of our general plan we take up the mean simultaneous departures of the temperature in these regions for which I have found observations to be readily available.   The results are given in Table X following.   In explaining them the object is to facilitate the work of using the departures, rather than to set forth in detail how they were formed.   The construction of the table is as follows.   The period under discussion, 1872–1900, is divided into periods during each of which the number of stations remain unchanged.   This is convenient because our general formulæ, as developed in Chapter I, involve a separate summation for each of these periods.

For the first period the entire United States is taken as a single region, because it is possible that, in the course of a month, a departure of temperature would have time to extend itself across the Rocky mountains from San Diego to Texas.   The mean departures found in the table are formed from the ten-day means given in the next chapter.   From and after 1874 the West Indian stations are combined with the United States, so as to form one general mean for all of North America.   The region South America is practically identical with the Argentine Republic.   The data for this region are also given in the ten-day tables.

It will be seen that the Indian stations and Batavia are treated as if completely independent.   Whether this is the case cannot be determined in advance of the general discussion.   The Australian departures are determined from an extended study and combination of the results given in the publications of the Adelaide Observatory by Sir Charles Todd.   For the most part they are formed from the mean of these six stations in which the departures were found to be least subject to fitful fluctuations

The departures at the several stations are numbered v_1, v_2, etc., in accordance with the system followed in Chapter I.   These index numbers are therefore the values of i in the equation of § 4–7.

Partly as a check, and partly to facilitate the ulterior discussion, the algebraic sum of the 12 departures for each year are found below the line for December.

The column Σ^2 which terminates the column for each year is the sum of the squares of all the departures for the year at each individual station.   From them the steadiness of the temperature may be inferred.

The mean τ, the general world departure so far as it can be inferred from the stations, and its square form the last two columns.   These enter into the formulæ of Chapter I, and are summed at the bottom of the columns.

TABLE X.

Monthly Simultaneous Deviations of Temperature in Widely Separated Regions.

FIRST PERIOD.

Date	U. S. v_1	S. Amer. v_2	India v_3	Batavia v_4	Mean τ	Mean τ^2	Date	U. S. v_1	S. Am. v_2	India v_3	Batavia v_4	Mean τ	Mean τ^2
1872							**1873**						
Jan.	− 0.4	+ 0.3	−0.4	− 1.3	−0.5	0.25	Jan.	+0.2	+0.9	−0.8	− 0.5	−0.1	0.01
Feb.	− 1.1	− 0.2	−0.2	− 1.1	−0.7	0.49	Feb.	−0.2	−0.6	+0.4	− 0.9	−0.3	0.09
Mar.	− 0.6	− 0.3	−0.2	− 0.3	−0.4	0.16	Mar.	+0.5	−1.0	−0.5	− 1.3	−0.6	0.36
April	+ 0.3	+ 0.1	−0.2	+ 0.2	+0.1	0.01	April	−0.1	−0.3	−0.2	− 0.3	−0.2	0.04
May	+ 0.4	− 0.6	−0.1	− 0.2	−0.1	0.01	May	+0.3	+1.5	−0.4	+ 0.1	+0.4	0.16
June	+ 1.2	+ 0.1	−0.3	− 0.2	+0.2	0.04	June	+0.4	+0.8	−0.1	− 0.2	+0.2	0.04
July	+ 0.7	− 0.1	−0.3	− 1.1	−0.2	0.04	July	+0.3	−0.1	+0.3	+ 0.1	+0.2	0.04
Aug.	+ 0.9	+ 0.5	−0.5	− 0.7	+0.1	0.01	Aug.	+0.3	+0.5	−0.1	+ 0.2	+0.2	0.04
Sept.	+ 0.5	+ 0.3	−0.2	+ 0.4	+0.2	0.04	Sept.	+0.4	+0.9	−0.2	+ 0.2	+0.3	0.09
Oct.	− 0.2	+ 0.9	−0.7	+ 0.2	+0.1	0.01	Oct.	−0.6	+0.3	−0.7	− 0.1	−0.3	0.09
Nov.	− 0.8	0.0	−0.2	− 0.7	−0.4	0.16	Nov.	−0.1	+0.5	−0.2	+ 0.3	+0.1	0.01
Dec.	− 1.4	+ 0.1	+0.7	− 0.2	−0.2	0.04	Dec.	+0.2	+0.5	−0.2	− 0.2	+0.1	0.01
Sum	− 0.7	+ 1.1	−2.6	− 5.0	−1.8	1.26	Sum	+1.6	+3.9	−2.7	− 2.6	0.0	0.98
Σ^2	+ 7.6	+ 1.7	+1.6	+ 5.4	....	......	Σ^2	+1.3	+6.6	+1.8	+ 2.9	....	......

SECOND PERIOD.

Date	N. Am. v_1	S. Am. v_2	India v_3	Batavia v_4	Mean τ	Mean τ^2	Date	N. Am. v_1	S. Am. v_2	India v_3	Bavaria v_4	Mean τ	Mean τ^2
1874							**1875**						
Jan.	+ 0.3	− 0.1	−0.8	− 0.4	−0.2	0.04	Jan.	0.0	−0.2	−0.4	− 0.8	−0.4	0.16
Feb.	+ 0.2	+ 0.2	−0.3	− 0.4	−0.1	0.01	Feb.	−0.2	−0.1	−0.3	− 0.7	−0.3	0.09
Mar.	+ 0.2	− 0.8	−0.4	− 0.3	−0.3	0.09	Mar.	−0.4	−1.3	+0.2	− 1.1	−0.6	0.36
April	− 0.5	0.0	−0.7	0.0	−0.3	0.09	April	−0.9	−0.8	+0.2	− 0.3	−0.4	0.16
May	+ 0.2	− 0.7	−0.8	− 0.8	−0.5	0.25	May	+0.4	−0.4	−0.2	− 0.2	−0.1	0.01
June	+ 0.2	− 0.1	−1.1	− 0.9	−0.5	0.25	June	+0.3	−1.5	+0.1	+ 0.6	−0.2	0.04
July	+ 0.4	− 0.9	−1.0	− 1.3	−0.7	0.49	July	−0.1	−1.0	+0.6	+ 0.3	−0.1	0.01
Aug.	− 0.1	+ 0.4	−0.1	− 1.1	−0.2	0.04	Aug.	+0.9	−0.9	−0.1	+ 0.3	+0.1	0.01
Sept.	− 0.6	0.0	−0.5	− 1.2	−0.6	0.36	Sept.	0.0	−0.1	−0.3	+ 0.7	+0.1	0.01
Oct.	+ 0.1	− 0.5	0.0	− 0.7	−0.3	0.09	Oct.	+0.8	−0.2	−0.7	− 0.4	−0.1	0.01
Nov.	+ 0.2	− 1.1	−0.2	− 0.8	−0.5	0.25	Nov.	+0.5	−0.1	+0.3	+ 0.1	+0.2	0.04
Dec.	+ 0.2	− 0.6	−0.3	− 0.4	−0.3	0.09	Dec.	+1.1	+0.4	+0.3	− 0.6	+0.3	0.09
Sum	+ 0.8	− 4.2	−6.2	− 8.3	−4.5	2.05	Sum	+2.4	−6.2	−0.3	− 2.1	−1.5	0.99
Σ^2	+ 0.9	+ 3.9	+4.5	+ 7.5	....	......	Σ^2	+4.1	+6.7	+1.5	+ 3.1	....	......
1876							**1877**						
Jan.	+ 1.4	− 0.5	+0.1	0.0	+0.3	0.09	Jan.	−0.1	−0.1	+0.4	− 0.2	0.0	0.00
Feb.	+ 1.1	− 0.7	−0.5	− 0.4	−0.1	0.01	Feb.	+0.3	−0.5	+0.6	− 0.7	−0.1	0.01
Mar.	− 0.5	+ 0.4	+0.4	+ 0.3	+0.2	0.04	Mar.	+0.4	+0.7	−0.2	− 0.4	+0.1	0.01
April	+ 0.3	− 0.5	0.0	+ 0.1	0.0	0.00	April	−0.2	+0.5	−0.4	+ 0.2	0.0	0.00
May	+ 0.1	+ 0.8	+0.2	+ 0.2	+0.3	0.09	May	−0.8	−0.1	−0.6	+ 1.3	0.0	0.00
June	+ 0.6	− 0.1	+0.1	− 0.2	+0.1	0.01	June	+0.5	+0.5	+0.2	+ 0.4	+0.4	0.16
July	+ 0.3	+ 0.9	+0.1	− 0.4	+0.2	0.04	July	+0.7	+2.1	+1.1	+ 0.2	+1.0	1.00
Aug.	− 0.3	− 0.8	+0.1	− 0.2	−0.3	0.09	Aug.	+0.6	+0.1	+0.9	− 0.4	+0.3	0.09
Sept.	− 0.1	+ 0.2	+0.2	− 0.6	−0.1	0.01	Sept.	+0.4	+0.1	+0.4	+ 0.2	+0.3	0.09
Oct.	− 0.5	− 0.4	+0.1	− 0.6	−0.3	0.09	Oct.	+0.3	+0.6	+0.5	+ 1.1	+0.6	0.36
Nov.	− 0.6	− 1.4	−0.4	− 0.3	−0.7	0.49	Nov.	−0.4	−0.7	+0.8	+ 2.8	+1.0	1.00
Dec.	− 1.9	− 1.2	−0.1	− 0.1	−0.8	0.64	Dec.	+0.2	+0.4	+1.4	+ 1.6	+0.9	0.81
Sum	− 0.1	− 3.3	+0.3	− 2.2	−1.2	1.60	Sum	+1.9	+5.0	+5.1	+ 6.1	+4.5	3.53
Σ^2	+ 8.3	+ 6.7	+0.6	+ 1.4	....	......	Σ^2	+2.5	+6.6	+6.2	+14.4	....	......

TABLE X.— *Continued.*

Monthly Simultaneous Deviations of Temperature in Widely Separated Regions.

SECOND PERIOD (*continued*).

Date	N. Am. v_1	S. Am. v_2	India v_3	Batavia v_4	Mean τ	τ^2	Date	N. Am. v_1	S. Am. v_2	India v_3	Batavia v_4	Mean τ	τ^2
1878							**1879**						
Jan.	− 0.6	− 0.6	+0.8	+ 2.2	+0.4	0.16	Jan.	−0.7	−0.4	+0.5	0.0	−0.2	0.04
Feb.	+ 0.4	+ 0.8	+1.1	+ 2.4	+1.2	1.44	Feb.	−0.4	−0.4	+0.4	+ 0.2	0.0	0.00
Mar.	+ 1.0	+ 1.3	+0.8	+ 1.8	+1.2	1.44	Mar.	+0.8	−0.6	+0.2	− 0.3	0.0	0.00
April	+ 0.9	+ 0.8	+0.2	+ 1.0	+0.7	0.49	April	0.0	−0.5	+0.1	0.0	−0.1	0.01
May	+ 0.4	− 0.3	+0.4	+ 1.0	+0.4	0.16	May	+0.3	−0.4	−0.8	− 0.3	−0.3	0.09
June	+ 0.8	+ 0.5	+1.1	+ 0.2	+0.6	0.36	June	−0.2	−0.9	−0.8	− 0.7	−0.7	0.49
July	+ 1.0	+ 0.6	−0.3	+ 1.1	+0.6	0.36	July	+0.4	+0.6	−0.2	− 0.1	+0.2	0.04
Aug.	+ 0.6	− 0.3	−0.1	+ 0.9	+0.3	0.09	Aug.	−0.1	−0.1	−0.5	− 1.0	−0.4	0.16
Sept.	− 0.1	+ 0.4	0.0	+ 1.3	+0.4	0.16	Sept.	+0.1	−0.9	+0.2	− 0.4	−0.3	0.09
Oct.	− 0.1	0.0	+0.7	+ 1.8	+0.6	0.36	Oct.	+0.5	−0.7	−0.6	− 0.8	−0.4	0.16
Nov.	− 0.2	+ 0.8	+0.8	+ 0.2	+0.4	0.16	Nov.	−0.2	+0.1	−0.9	0.0	−0.2	0.04
Dec.	− 1.9	− 0.6	+0.1	0.0	−0.6	0.36	Dec.	+0.1	−0.7	−1.1	+ 0.4	−0.3	0.09
Sum	+ 2.2	+ 3.4	+5.6	+13.9	+6.2	5.54	Sum	+0.6	−4.9	−3.6	− 3.0	−2.7	1.21
Σ²	+ 8.2	+ 5.3	+5.0	+22.7	....	......	Σ²	+1.8	+4.2	+4.4	+ 2.7	....	......
1880							**1881**						
Jan.	+ 1.8	− 0.7	−0.5	− 0.8	−0.1	0.01	Jan.	−0.7	0.0	+0.7	− 0.9	−0.2	0.04
Feb.	+ 1.8	− 0.5	−0.6	+ 0.4	+0.3	0.09	Feb.	+0.2	+0.5	+0.4	+ 0.6	+0.4	0.16
Mar.	− 0.4	− 0.3	+0.4	− 0.4	−0.2	0.04	Mar.	−1.0	+1.2	+0.2	− 0.2	+0.1	0.01
April	+ 0.1	+ 0.2	+0.5	− 0.4	+0.1	0.01	April	0.0	−0.1	+0.1	+ 0.4	+0.1	0.01
May	+ 0.1	+ 0.9	+0.2	− 0.2	+0.2	0.04	May	+0.9	+0.5	+0.6	+ 0.6	+0.6	0.36
June	+ 0.4	+ 1.9	+0.2	− 1.4	+0.3	0.09	June	+0.9	+0.2	+0.1	+ 0.3	+0.4	0.16
July	+ 0.4	+ 1.0	−0.2	− 1.1	0.0	0.00	July	+0.2	−0.4	+0.6	+ 0.2	+0.2	0.04
Aug.	− 0.3	+ 1.9	0.0	− 0.7	+0.2	0.04	Aug.	0.0	−0.4	0.0	0.0	−0.1	0.01
Sept.	− 0.2	− 0.7	−0.5	− 0.9	−0.6	0.36	Sept.	0.0	+0.1	−0.2	0.0	0.0	0.00
Oct.	− 0.5	− 1.0	0.0	− 0.7	−0.5	0.25	Oct.	+0.3	+0.2	+0.3	+ 0.9	+0.4	0.16
Nov.	− 1.6	+ 0.3	+0.2	− 0.6	−0.4	0.16	Nov.	−0.3	+0.1	−0.1	+ 0.1	−0.1	0.01
Dec.	− 0.8	+ 0.7	+0.1	− 0.6	−0.2	0.04	Dec.	+0.7	+0.6	+0.1	+ 0.9	+0.6	0.36
Sum	+ 0.8	+ 3.7	−0.2	− 7.4	−0.9	1.13	Sum	+1.2	+2.5	+2.8	+ 2.9	+2.4	1.32
Σ²	+10.5	+11.9	+1.2	+ 7.0	....	......	Σ²	+3.8	+2.6	+1.6	+ 3.5	....	......
1882													
Jan.	− 0.2	0.0	+0.5	+ 0.1	+0.1	0.01							
Feb.	+ 0.5	− 0.6	+0.1	+ 0.7	+0.2	0.04							
Mar.	+ 0.6	− 0.2	+0.3	− 0.3	+0.1	0.01							
April	+ 0.3	− 1.1	+0.1	+ 0.1	−0.1	0.01							
May	0.0	0.0	−0.2	− 0.9	−0.3	0.09							
June	+ 0.7	− 0.2	+0.1	− 1.0	−0.1	0.01							
July	0.0	− 0.8	−0.7	− 1.0	−0.4	0.16							
Aug.	+ 0.1	+ 0.3	0.0	− 0.1	+0.1	0.01							
Sept.	0.0	− 0.2	0.0	− 1.0	−0.3	0.09							
Oct.	+ 0.2	+ 1.1	−0.5	− 1.3	−0.1	0.01							
Nov.	− 0.7	− 0.2	−0.4	− 0.9	−0.6	0.36							
Dec.	− 0.7	− 1.4	−0.2	+ 0.3	−0.5	0.25							
Sum	+ 0.8	− 3.3	−0.9	− 4.4	−1.9	1.05							
Σ²	+ 2.2	+ 5.5	+1.2	+ 6.0	....	......							

TABLE X.—*Continued.*

Monthly Simultaneous Deviations of Temperature in Widely Separated Regions.

THIRD PERIOD.

Date	N. Am. v_1	S. Am. v_2	India v_3	Batavia v_4	Australia v_5	Mean τ	τ^2	Date	N. Am. v_1	S. Am. v_2	India v_3	Batavia v_4	Australia v_5	Mean τ	τ^2
1883								**1884**							
Jan.	−0.3	−0.1	+0.6	− 0.1	+0.4	+0.1	0.01	Jan.	−0.7	+0.2	−0.9	−0.4	−0.8	+0.1	0.01
Feb.	+0.3	+0.1	−0.2	− 0.4	−1.1	−0.3	0.09	Feb.	+0.6	−0.3	−0.7	−0.6	+0.1	−0.3	0.09
Mar.	+0.3	+0.7	−0.2	+ 0.4	+0.2	+0.3	0.09	Mar.	+0.5	+0.7	−0.3	−0.9	+0.6	+0.3	0.09
April	−0.1	−0.4	−0.1	+ 0.5	+1.3	0.0	0.00	April	−0.5	−0.4	−0.6	−0.4	−0.2	0.0	0.00
May	−0.2	+0.3	+0.2	− 0.3	−0.4	−0.1	0.01	May	+0.4	−0.9	−0.4	−0.5	+0.1	−0.1	0.01
June	+0.7	+0.9	−0.2	+ 0.9	+1.4	+0.7	0.49	June	−0.1	−0.8	+0.3	+0.1	+0.4	+0.7	0.49
July	+0.1	+0.1	−0.5	+ 0.4	+0.8	+0.2	0.04	July	+0.1	−0.1	+0.1	−0.6	−0.7	+0.2	0.04
Aug.	0.0	−1.0	+0.2	− 0.1	0.0	−0.2	0.04	Aug.	−0.3	+2.4	+0.4	0.0	+1.2	−0.2	0.04
Sept.	+0.4	−0.2	+0.1	− 0.3	−1.0	−0.2	0.04	Sept.	−0.4	−0.4	−0.3	+0.1	+0.2	−0.2	0.04
Oct.	−0.3	+0.2	−0.8	− 0.6	−0.8	−0.5	0.25	Oct.	+0.5	−0.3	−0.5	0.0	−0.3	−0.5	0.25
Nov.	+0.1	+0.3	−0.7	− 1.1	−0.3	−0.3	0.09	Nov.	+0.1	+0.1	−1.0	−0.3	−0.1	−0.3	0.29
Dec.	+0.3	+0.2	−1.4	− 0.7	−0.7	−0.5	0.25	Dec.	+0.8	0.0	−0.5	−0.7	−1.4	−0.5	0.25
Sum.	+1.3	+1.1	−3.0	− 2.4	−0.2	−0.8	1.40	Sum	+1.0	+0.2	−4.4	−4.2	−0.9	−0.8	1.40
Σ^2	+1.2	+2.7	+3.7	+ 3.9	+8.1	....	......	Σ^2	+2.6	+8.3	+3.8	+2.8	+5.2	....	......
1885								**1886**							
Jan.	+0.4	+0.5	−0.1	− 0.8	−0.4	−0.1	0.01	Jan.	−1.1	+0.6	−0.2	+0.9	+1.1	+0.3	0.09
Feb.	−0.2	−0.1	−0.8	− 0.8	−1.1	−0.6	0.36	Feb.	0.0	−0.3	−0.3	−0.1	−1.3	−0.4	0.16
Mar.	−0.1	−0.6	−0.4	− 1.0	−1.4	−0.7	0.49	Mar.	−0.6	+0.7	+0.2	0.0	0.0	+0.1	0.01
April	+1.0	+1.4	−0.9	− 0.2	−0.2	+0.2	0.04	April	−0.8	+0.1	−0.2	0.0	+0.1	−0.2	0.04
May	+0.3	+0.4	−0.8	− 0.4	+1.4	+0.2	0.04	May	+0.2	−0.4	−0.2	+0.2	0.0	0.0	0.00
June	0.0	−0.3	0.0	+ 0.5	−0.9	−0.1	0.01	June	+0.5	−0.6	−0.8	+0.2	−0.3	−0.2	0.04
July	+0.7	−0.3	+0.4	+ 0.6	+0.2	+0.3	0.09	July	+0.2	−1.0	−0.5	+0.9	+0.7	+0.1	0.01
Aug.	+0.4	−0.6	+0.3	− 0.1	+0.8	+0.2	0.04	Aug.	+0.4	−0.9	−0.2	+0.3	+0.3	0.0	0.00
Sept.	+0.2	+0.8	+0.3	+ 0.5	0.0	+0.4	0.16	Sept.	+0.5	−0.5	+0.3	+0.5	+1.4	+0.4	0.16
Oct.	−0.2	+0.2	+0.3	+ 0.3	+0.9	+0.3	0.09	Oct.	−0.4	−0.7	−0.1	−0.2	−1.3	−0.5	0.25
Nov.	+0.1	+1.1	+0.1	+ 0.1	−0.1	+0.3	0.09	Nov.	−1.1	−0.1	+0.3	+0.6	+0.4	0.0	0.00
Dec.	−0.2	+0.1	−0.1	+ 0.4	+0.7	+0.2	0.04	Dec.	−0.3	+0.3	−0.1	−0.5	−0.3	−0.2	0.04
Sum	+2.4	+2.6	−1.5	− 0.9	−0.1	+0.6	1.46	Sum	−2.5	−2.8	−1.8	+2.8	+0.8	−0.6	0.84
Σ^2	+2.0	+5.2	+2.8	+ 3.5	+8.1	....	......	Σ^2	+4.3	+4.2	+1.1	+2.5	+7.6	....	......
1887								**1888**							
Jan.	0.0	+0.6	−0.3	+10.4	+1.1	+0.4	0.16	Jan.	−0.9	−0.2	0.0	−1.0	+0.2	−0.4	0.16
Feb.	+1.3	−0.3	−0.5	− 0.1	−0.2	+0.1	0.01	Feb.	+1.1	+0.4	+0.3	−0.5	−1.0	+0.1	0.01
Mar.	+0.5	−0.3	−0.3	− 0.4	−0.2	−0.1	0.01	Mar.	−0.7	+0.3	+0.3	+0.2	−0.8	−0.1	0.01
April	−0.1	−0.6	−0.4	− 0.4	−0.1	−0.3	0.09	April	+0.8	+0.1	+0.3	−0.2	+1.0	+0.4	0.16
May	+0.6	−0.5	+0.3	− 1.1	−0.8	−0.3	0.09	May	−0.2	0.0	−0.3	−0.3	+0.5	−0.1	0.01
June	−0.4	+1.6	−0.8	− 1.0	−0.9	−0.3	0.09	June	+0.2	−1.2	+0.1	+0.9	+0.8	+0.2	0.04
July	+0.2	−0.3	−0.2	− 0.6	0.0	−0.2	0.04	July	0.0	+0.9	−0.1	+0.4	+0.4	+0.2	0.09
Aug.	−0.1	+2.1	−0.6	− 0.2	−0.3	−0.2	0.04	Aug.	+0.3	+0.7	−0.3	+0.7	−0.4	+0.3	0.04
Sept.	0.0	−0.3	−0.7	− 1.0	−0.9	+0.6	0.36	Sept.	+0.3	+1.0	+0.3	+0.7	+0.9	+0.6	0.36
Oct.	−0.1	0.0	−0.3	− 0.7	−0.2	−0.3	0.09	Oct.	+0.1	+1.1	+0.4	+1.1	+0.3	+0.6	0.36
Nov.	+0.2	−0.7	+0.2	− 0.9	−1.3	−0.5	0.25	Nov.	0.0	+0.6	+0.8	+1.2	+1.8	+0.9	0.81
Dec.	−0.4	+0.2	0.0	− 0.4	−0.1	−0.1	0.01	Dec.	−0.1	+1.2	+0.1	+1.2	+1.2	+0.7	0.49
Sum	+1.7	+1.5	−3.6	− 6.4	−3.9	−2.0	1.24	Sum	+0.9	+4.9	+1.9	+4.4	+4.9	+3.4	2.54
Σ^2	+2.7	+8.9	+2.3	+ 5.7	+5.2	....	......	Σ^2	+3.3	+7.0	+1.4	+7.3	+9.3	....	......
1889															
Jan.	+0.3	+1.0	+0.3	+ 1.6	+1.1	+0.9	0.81								
Feb.	−0.3	+0.5	+0.3	+ 1.5	+0.6	+0.5	0.25								
Mar.	+0.3	+1.1	+0.3	+ 1.0	+1.4	+0.8	0.64								
April	+0.7	+0.3	+0.1	+ 1.5	+0.7	+0.7	0.49								
May	+0.1	+0.6	+0.4	+ 1.2	+0.4	+0.5	0.25								
June	+0.1	−0.5	+0.1	+ 0.3	+1.0	+0.2	0.04								
July	+0.4	+0.3	−0.1	+ 0.3	0.0	+0.2	0.04								
Aug.	+0.1	−0.7	+0.3	+ 0.9	+0.1	+0.1	0.01								
Sept.	0.0	−1.0	+0.2	+ 0.3	−0.1	−0.1	0.01								
Oct.	−0.2	+0.4	−0.2	− 0.6	+0.8	0.0	0.00								
Nov.	+0.6	+0.3	−0.7	+ 0.1	+0.5	+0.2	0.04								
Dec.	+1.9	+0.8	−0.4	+ 0.8	−0.3	+0.6	0.36								
Sum	+4.0	+3.1	+0.6	+ 8.9	+6.2	+4.6	2.94								
Σ^2	+5.0	+5.6	+1.3	+11.5	+6.2	....	......								

TABLE X.— *Continued.*

Monthly Simultaneous Deviations of Temperature in Widely Separated Regions.

FOURTH PERIOD.

Date	N. Am. v_1	S. Am. v_2	India v_3	Batavia v_4	Apia v_5	Australia v_6	τ	τ^2	Date	N. Am. v_1	S. Am. v_2	India v_3	Batavia v_4	Apia v_5	Australia v_6	τ	τ^2
1890									**1891**								
Jan.	+1.4	0.0	+0.1	+1.5	−0.3	+1.9	+0.8	0.64	Jan.	0.0	+ 0.3	+0.1	+ 0.4	+0.4	−0.7	+0.1	0.01
Feb.	+1.4	+ 0.1	+0.5	+0.1	0.0	+0.4	+0.4	0.16	Feb.	+0.6	+ 1.4	−0.7	− 0.1	+0.1	−0.8	+0.1	0.01
Mar.	−0.4	− 0.9	+0.6	+0.2	+0.5	+0.5	+0.1	0.01	Mar.	−0.6	+ 1.2	−0.4	− 0.3	+0.2	+0.6	+0.1	0.01
April	+0.1	+ 0.3	+0.1	+0.4	−0.4	+0.3	+0.1	0.01	April	−0.4	+ 0.4	−0.5	0.0	0.0	−0.3	−0.1	0.01
May	+0.3	− 0.3	−0.1	−0.3	+0.1	+0.4	0.0	0.00	May	−0.4	− 0.8	−0.5	+ 1.0	+0.1	+0.8	+0.1	0.01
June	+0.1	− 1.7	−1.1	−0.8	+0.1	+1.1	−0.4	0.16	June	+0.2	+ 0.7	+1.0	+ 0.7	+0.1	−0.7	+0.3	0.09
July	+0.3	+ 0.2	−0.8	−0.9	−0.5	−0.4	−0.3	0.09	July	−0.1	− 0.4	0.0	+ 0.2	−0.5	−0.3	−0.2	0.04
Aug.	−0.5	− 1.0	−0.9	−0.7	−0.2	−0.3	−0.6	0.36	Aug.	+0.1	− 0.3	+0.3	− 0.7	−0.4	−0.1	−0.2	0.04
Sept.	−0.5	− 0.5	−0.4	−0.8	+0.6	+0.9	−0.1	0.01	Sept.	+0.5	− 0.3	−0.1	+ 0.5	0.0	+0.1	+0.1	0.01
Oct.	−0.1	+ 0.1	−0.1	−1.3	+0.2	−0.3	−0.3	0.09	Oct.	−0.6	+ 1.0	+0.3	+ 1.8	+0.5	0.0	+0.5	0.25
Nov.	+0.9	+ 0.6	+0.1	−1.7	+0.1	−1.2	−0.2	0.04	Nov.	+0.1	− 0.1	+0.2	+ 1.4	+0.1	−0.4	+0.2	0.04
D. c.	+0.5	+ 0.8	+0.4	−0.3	+0.4	−0.5	+0.2	0.04	Dec.	+0.2	− 1.3	+0.3	+ 0.6	+0.2	−0.4	−0.1	0.01
Sum	+3.5	− 2.3	−1.6	−4.6	+0.6	+2.8	−0.3	1.61	Sum	−0.4	+ 1.8	+0.1	+ 5.5	+0.8	−2.2	−0.9	0.53
Σ^2	+5.8	+ 6.1	+3.6	+9.7	+1.3	+8.3	......	......	Σ^2	+1.8	+ 7.9	+2.4	+ 8.1	+0.8	+3.2	......	......
1892									**1893**								
Jan.	−0.9	0.0	+0.9	+0.3	−0.3	−0.3	−0.1	0.01	Jan.	−1.3	0.0	−0.6	− 0.8	−0.7	−0.4	−0.6	0.36
Feb.	+0.2	+ 1.6	+1.4	+1.3	+0.4	+0.7	+0.9	0.81	Feb.	+0.1	− 1.7	−1.4	− 0.6	−0.5	+0.1	−0.5	0.25
Mar.	−1.0	− 1.1	+0.3	+0.2	+0.7	+0.8	0.0	0.00	Mar.	−0.3	− 0.1	−0.7	+ 0.1	−0.2	+0.9	−0.1	0.01
April	−0.3	− 0.7	+1.7	−0.7	+0.2	−1.0	−0.1	0.01	April	+0.1	− 0.6	+0.4	− 0.4	−0.8	−0.3	−0.3	0.09
May	−0.2	− 1.4	+0.7	−0.1	−0.4	−0.2	−0.3	0.09	May	−0.2	− 1.4	−0.4	− 0.3	−0.6	+1.2	−0.3	0.09
June	−0.5	− 2.2	−0.3	+0.1	+0.2	+0.4	−0.4	0.16	June	−0.1	− 3.3	−0.7	− 0.8	+0.3	−0.4	−0.8	0.64
July	−0.7	− 0.3	+0.3	+0.2	−0.1	+0.4	0.0	0.00	July	+0.1	+ 0.1	−0.2	− 0.6	−0.1	+0.8	0.0	0.00
Aug.	0.0	− 2.3	−0.5	−0.2	+0.2	+0.8	−0.3	0.09	Aug.	−0.1	− 2.6	0.0	− 0.4	−0.1	+0.4	−0.5	0.25
Sept.	+0.2	− 1.2	−0.7	−0.7	−0.3	−0.1	−0.5	0.25	Sept.	−0.1	− 3.0	−0.1	− 0.3	−0.3	+0.3	−0.6	0.36
Oct.	−0.3	+ 0.2	+0.2	0.0	+0.1	−0.2	0.0	0.00	Oct.	−0.1	− 2.3	−0.4	− 0.5	−0.5	+0.5	−0.6	0.36
Nov.	−0.2	− 0.2	−0.5	+0.4	−0.7	+0.7	−0.1	0.01	Nov.	−0.2	− 2.1	+0.2	− 0.5	0.0	−0.6	−0.5	0.25
Dec.	−0.5	− 1.0	−0.2	+0.5	−0.4	−1.2	−0.5	0.25	Dec.	+0.8	+ 1.1	+0.2	− 0.9	−0.3	0.0	+0.2	0.04
Sum	−4.2	− 8.6	+3.3	+1.3	−0.4	+0.8	−1.4	1.68	Sum	−1.3	−15.9	−3.7	− 6.0	−3.8	+2.5	−4.6	2.70
Σ^2	+2.9	+18.9	7.4	+3.2	+1.8	+5.1	......	......	Σ^2	+2.4	+42.9	+4.0	+ 3.8	+2.2	+4.2	......	......
1894									**1895**								
Jan.	+0.7	+ 0.3	+0.7	−0.9	−0.2	−0.4	0.0	0.00	Jan.	−0.1	− 1.8	−0.4	− 0.3	−0.1	−0.8	−0.6	0.36
Feb.	−0.9	+ 0.4	+1.1	−0.5	−0.4	−1.7	−0.3	0.09	Feb.	−2.6	+ 0.6	+0.1	− 0.5	−1.0	−0.3	−0.6	0.36
Mar.	+0.1	− 1.4	+0.6	−0.4	−0.2	+0.1	−0.2	0.04	Mar.	−0.2	+ 1.5	+0.4	− 0.2	−0.1	−0.3	+0.2	0.04
April	−0.1	0.0	+0.3	−0.2	+0.5	+0.1	+0.1	0.01	April	−0.3	+ 1.5	−0.1	+ 0.3	+0.2	−0.2	+0.2	0.04
May	0.0	+ 2.0	+0.1	−0.8	−0.1	−0.9	+0.1	0.01	May	+0.2	+ 1.2	−0.2	− 0.2	+0.8	−0.4	+0.2	0.04
June	−0.8	− 0.7	−0.1	−0.4	+0.3	−0.3	−0.3	0.09	June	−0.1	+ 3.3	0.0	+ 0.1	+0.2	+0.3	+0.6	0.36
July	−0.3	− 0.1	−0.1	+0.4	−0.3	−0.1	−0.1	0.01	July	−0.3	+ 2.0	+0.1	− 0.8	+0.4	−0.2	+0.2	0.04
Aug.	−0.1	− 0.6	+0.3	+0.3	+0.6	−0.2	+0.1	0.01	Aug.	0.0	+ 1.5	−0.1	− 0.3	−0.2	+0.4	+0.2	0.04
Sept.	−0.3	− 1.2	−0.5	+0.1	+0.5	−1.4	−0.5	0.25	Sept.	+0.6	− 0.7	−0.3	+ 0.9	−0.4	−0.2	0.0	0.00
Oct.	−0.1	− 1.8	−0.4	0.0	+0.6	−0.5	−0.4	0.16	Oct.	−0.2	− 0.9	+0.2	+ 1.0	.0.0	+1.2	+0.2	0.04
Nov.	+0.2	+ 0.4	−0.5	0.0	−0.4	−0.1	−0.1	0.01	Nov.	−0.1	− 1.2	+0.9	+ 0.4	−0.4	−0.7	−0.2	0.04
Dec.	−0.1	− 0.2	−0.3	0.0	−0.1	−0.4	−0.2	0.04	Dec.	−1.2	+ 0.6	−0.7	+ 0.4	+0.4	−0.2	0.0	0.00
Sum	−1.7	− 2.9	+1.2	−2.4	+0.8	−5.8	−1.8	0.66	Sum	−4.3	+ 7.6	+0.8	+ 0.8	−0.2	−1.4	+0.4	1.36
Σ^2	+2.1	+12.0	+3.0	+2.3	+1.8	+6.4	......	......	Σ^2	+8.8	+28.6	+1.3	+ 3.3	+2.4	+3.2	......	......
1896									**1897**								
Jan.	+0.2	− 0.2	+1.1	+0.6	−0.2	+0.2	+0.3	0.09	Jan.	−0.4	− 0.5	−0.2	+ 2.7	−0.3	−1.0	−0.1	0.01
Feb.	+0.2	− 0.2	+0.3	+0.5	−0.4	−0.5	0.0	0.00	Feb.	+0.6	+ 0.4	−0.3	+ 1.6	−0.2	+0.4	+0.6	0.36
Mar.	−0.6	+ 0.4	+0.5	+0.4	+0.1	+0.2	+0.2	0.04	Mar.	+0.1	+ 2.7	−0.8	+ 1.0	+0.3	−1.2	+0.4	0.16
April	−0.5	+ 1.1	+1.2	−0.3	−0.2	−0.1	+0.2	0.04	April	0.0	+ 1.9	+0.2	+ 0.2	+0.1	+0.8	+0.5	0.25
May	+0.1	0.0	+0.8	+0.3	+0.6	−0.6	+0.2	0.04	May	+0.1	+ 1.6	+0.1	+ 1.3	−0.1	−0.3	+0.4	0.16
June	+0.7	− 0.2	+0.2	+0.6	−0.4	−1.6	−0.1	0.01	June	−0.3	+ 0.7	+1.1	+ 1.3	+0.3	+0.9	+0.7	0.49
July	0.0	+ 3.5	+0.1	+0.6	−1.1	−0.8	+0.4	0.16	July	−0.2	− 1.8	+0.3	+ 0.9	−0.1	+1.2	0.0	0.00
Aug.	+0.4	+ 3.9	−0.3	+0.5	0.0	−0.9	+0.6	0.36	Aug.	0.0	− 1.8	+0.6	+ 0.9	+0.2	−0.4	−0.1	0.01
Sept.	+0.1	+ 2.4	+0.9	+0.8	+0.3	−0.9	+0.6	0.36	Sept.	+0.1	− 1.0	+0.5	+ 1.2	−0.5	+0.2	+0.1	0.01
Oct.	+0.3	+ 0.7	+1.4	+1.4	+0.6	+0.5	+0.8	0.64	Oct.	+0.9	+ 1.0	+0.3	+ 0.5	−0.2	−0.3	+0.4	0.16
Nov.	+1.1	+ 0.3	+1.2	+1.8	−0.4	+0.4	+0.7	0.49	Nov.	+0.9	+ 0.6	−0.6	+ 0.4	−0.7	+0.9	0.0	0.00
Dec.	+0.5	− 0.4	+0.9	+0.1	+0.5	−0.3	+0.2	0.04	Dec.	+0.5	+ 0.8	−0.7	+ 1.0	+0.1	+0.7	+0.6	0.36
Sum	+2.5	+11.3	+8.3	+7.3	−0.6	−4.4	+4.1	2.27	Sum	+2.3	+ 3.4	+0.5	+13.0	−1.1	+1.9	+3.5	1.97
Σ^2	+2.8	+35.4	+8.6	+7.8	+3.9	+5.9	......	......	Σ^2	+2.5	+23.8	+3.6	+18.7	+1.0	+7.1	......	......

TABLE X.—*Concluded.*

Monthly Simultaneous Deviations of Temperature in Widely Separated Regions.

FOURTH PERIOD (*concluded*).

Date	N. Am. v_1	S. Am. v_2	India v_3	Batavia v_4	Apia v_5	Australia v_6	Mean τ	τ^2
1898								
Jan.	+0.3	+ 0.7	+0.4	+0.2	−0.6	+0.7	+0.3	0.09
Feb.	+0.2	+ 1.4	+0.2	+0.1	−0.6	+1.4	+0.4	0.16
Mar.	−0.1	− 0.8	+0.4	+0.3	−0.8	−0.1	−0.2	0.04
April	0.0	− 1.2	+0.6	+0.4	−0.3	−0.7	−0.2	0.04
May	−0.5	+ 1.4	+0.4	+0.6	−0.3	−1.5	−0.2	0.04
June	+0.1	+ 2.4	0.0	+0.5	−0.1	−0.1	+0.5	0.25
July	+0.2	− 1.5	+0.1	+0.3	+0.8	+0.3	0.0	0.00
Aug.	+0.2	− 2.1	+0.3	+0.7	0.0	+0.8	0.0	0.00
Sept.	+0.7	− 1.2	0.0	+0.4	+0.2	+0.3	+0.1	0.01
Oct.	0.0	− 2.5	+1.4	−0.6	−0.2	+0.5	−0.2	0.04
Nov.	0.0	− 2.2	+1.2	−0.3	−0.6	−1.1	−0.5	0.25
Dec.	−0.5	− 0.1	+1.0	−0.2	−0.1	+0.9	+0.1	0.01
Sum	−0.6	− 5.7	+6.0	+2.4	−2.6	+1.4	+0.1	0.93
Σ^2	+1.0	+31.3	+5.5	+2.2	+2.6	+8.2	....	......

FIFTH PERIOD.

Date	N. Am. v_1	India v_2	Batavia v_3	Apia v_4	Australia v_5	Mean τ	τ^2	Date	N. Am. v_1	India v_2	Batavia v_3	Apia v_4	Australia v_5	Mean τ	τ^2
1899								**1900**							
Jan.	+1.0	−1.1	−0.5	−0.8	− 2.6	−0.8	0.64	Jan.	+1.0	−0.4	+0.7	+0.5	+0.3	+0.4	0.16
Feb.	−1.2	+0.2	−0.6	−0.2	+ 1.3	−0.1	0.01	Feb.	−0.2	−0.3	+0.7	+0.5	+0.7	+0.2	0.04
Mar.	0.0	+0.5	−0.4	−0.2	+ 0.7	+0.1	0.01	Mar.	+0.4	+0.1	+0.7	−0.3	−0.6	+0.1	0.01
April	−0.4	0.2	− 0.3	−0.1	+ 0.3	−0.1	0.01	April	−0.3	−0.1	+0.4	0.0	−0.5	−0.1	0.01
May	−0.6	+0.3	+0.4	−0.2	− 0.7	−0.2	0.04	May	+0.2	+0.2	+0.1	−0.2	+0.2	+0.1	0.01
June	−0.4	−0.3	−0.3	−0.6	− 0.2	−0.4	0.16	June	+0.2	+1.1	+0.5	−0.6	+0.2	+0.3	0.09
July	+0.1	+0.6	+0.6	+0.3	− 1.3	+0.1	0.01	July	0.0	+0.9	+0.6	−0.4	−0.2	+0.2	0.04
Aug.	−0.2	+0.8	+0.1	−0.1	− 0.6	0.0	0.00	Aug.	+0.3	+0.2	−0.8	+0.3	−0.2	0.0	0.00
Sept.	+0.6	+0.7	+0.1	+0.4	− 0.2	+0.3	0.09	Sept.	+0.4	+0.4	+1.0	+0.5	−0.7	+0.3	0.09
Oct.	+0.2	+1.0	+0.4	+0.1	− 0.7	+0.2	0.04	Oct.	+1.1	0.0	+1.0	+0.3	+0.1	+0.5	0.25
Nov.	+0.9	+0.3	+0.4	+0.4	− 0.4	+0.3	0.09	Nov.	+1.5	+0.6	+0.6	−0.1	+0.3	+0.6	0.36
Dec.	0.0	+0.6	−0.1	−0.1	+ 0.2	+0.1	0.01	Dec.	+1.1	+1.1	+1.1	−0.2	−0.2	+0.6	0.36
Sum	0.0	+4.0	−0.2	−1.1	− 4.2	−0.5	1.11	Sum	+5.7	+3.8	+6.6	+0.3	−0.4	+3.2	1.42
Σ^2	+4.4	+4.6	+2.0	+1.5	+12.4	....	......	Σ^2	+6.2	+4.1	+6.5	+1.5	+1.8	....	......

To investigate the correlation among the stations we apply the method and formulæ of § 4, as we have done in the case of the annual deviations. For example, we have for the first period, 1871 and 1872,

$$1871: \quad \Sigma_i v_i^2 = 7.6 + 1.7 + 1.6 + 5.4 = 16.3$$

$$1872: \quad \text{“} \quad = 1.3 + 6.6 + 1.8 + 2.9 = 12.6$$

also

$$\Sigma_j \tau^2 = 1.26 + 0.98 = 2.24$$

Thus, this period alone gives

$$\Sigma_{i,j} v^2 = 28.9$$

Since $n = 4$, and r, the number of monthly terms, is 24,

$$n^2 \Sigma_j \tau^2 = 35.8$$

Thus (9) gives the equation

$$288\tau_0^2 = \Delta = +6.9$$

Carrying this computation through all the time-terms we have the following results :

Period	n	r	$\dfrac{\Sigma v^2}{n}$	$n\Sigma \tau^2$	Equation for τ_0^2
1872–73	4	21	7.2	9.0	$288\tau_0^2 = + \quad 7$
1874–82	4	108	47.6	73.7	1296 $\quad +104$
1883–89	5	84	33.3	58.6	1680 $\quad +127$
1890–98	6	108	67.5	82.3	3240 $\quad + 89$
1899–00	5	24	9.0	12.7	480 $\quad + 18$
Sum	.	......	164.6	236.3	$6984\tau_0^2 = +345$

A positive correlation is well shown, leading to the mean result

$$\tau_0^2 = .0493$$

$$\tau_0 = \pm 0.22° \text{ C.} = \pm 0.4° \text{ Fahr.}$$

When we add in the equation from Dove's work the final equation is

$$9396\tau_0^2 = 401$$

whence

$$\tau_0 = \pm 0°.21$$

The existence of the positive correlation is beyond serious question, but before we accept it as cosmical, we must learn whether it holds between the more distant stations, as well as between those in neighboring great geographic zones.

As no correlation but a cosmical one can exist between the North American and the other regions, we first compare that with the others. The table shows that simultaneous temperatures in North and South America are available from 1872 to 1898, a period of 324 months. Forming the sum of the 324 products $v_1 v_2$ we find the result

$$\Sigma vv' = \Delta = +15.3$$

Proceeding in the same way with the other stations the collected results are :

North America — South America ;	$r = 324$	$\Sigma vv = + 15.3$
" " — India ;	" 348	" $- 2.8$
" " — Batavia ;	" 348	" $+ 18.4$
" " — Australia ;	" 216	" 0.0
" " — Apia ;	" 132	" $+ 1.0$
Sum		$+ 31.9$

The South American products being formed in the same way, the results of their summation are:

$$
\begin{array}{llll}
\text{South America — India;} & r = 348 & \Sigma vv' = & + 18. \\
\text{" \quad " \quad — Australia;} & \text{" } 192 & \text{" } & - 5. \\
\text{" \quad " \quad — Apia;} & \text{" } 108 & \text{" } & + 4. \\
\text{" \quad " \quad — Batavia;} & \text{" } 327 & \text{" } & + 36. \\
\text{Sum} & & & + 53.
\end{array}
$$

Next we have

$$
\begin{array}{llll}
\text{India — Batavia;} & r = 348 & \Sigma vv' = & + 51. \\
\text{" \quad — Apia;} & \text{" } 132 & \text{" } & 0. \\
\text{" \quad — Australia;} & \text{" } 216 & \text{" } & + 5. \\
\text{Sum} & & & + 56.
\end{array}
$$

Then

$$
\begin{array}{llll}
\text{Batavia — Australia;} & r = 216 & \Sigma vv' = & + 26. \\
\text{" \quad — Apia;} & \text{" } 132 & \text{" } & + 2. \\
\text{Sum} & & & + 28.
\end{array}
$$

$$\Sigma vv' = + 4$$

It will be seen that, while there seems to be a general tendency toward a positive correlation, the largest part of Δ arises from the two combinations India-Batavia and Batavia-Australia. These pairs being in comparative geographic proximity, we may well throw them out. The remaining pairs give:

Whole number of products, 2924
Sum of all these products + 96

Hence,

$$
\begin{aligned}
\text{Mean } vv' &= \text{mean } \tau_0^2 = + 0.033 \\
\text{Mean } \tau_0 &= \pm 0°.18 \text{ C.} = \pm 0°.32 \text{ Fahr.}
\end{aligned}
$$

It therefore seems that the monthly departures of temperature indicate fluctuations in the general world temperature of which the general amount is about $\pm 0°.18$ C. on each side of the normal mean value   This is scarcely greater than the degree of correlation which we should expect to be shown from our omission to correct the normal tables for the sun-spot inequality, and from the systematic deviations of the annual temperature brought out in § 9. The evidence is therefore rather weak in favor of very minute fluctuations in the sun's radiation for periods greater than one month and less than several years. If they exist, they are too small to produce any noticeable meteorological effect.

CHAPTER V.

STUDY OF TEN-DAY TERMS.

§ 13. *Stations and Material Used.*

The term of ten days was chosen because it has been extensively adopted, especially in the *Dekadenberichte* of the German *Seewarte*. Mean temperatures for this purpose being available in a number of cases, the labor of forming them for the entire work was not necessary. A term of one fourth or one fifth the sun's rotation would have been better adapted to bringing out fluctuations having the period of that rotation ; but a lesser period than ten days would be subject to the drawback that small fluctuations in the radiation require time to produce their full effect upon the temperature, so that little indication of their effect could be expected.

Strictly speaking, the period is not ten days but one third of a month. When it was necessary to form independent mean temperatures from daily records, the year was divided into thirty-six parts as nearly equal as possible. There were, therefore, thirty or thirty-one periods of ten days each, and five or six of eleven days in each year. But when the ten-day means had been taken on a different system, the month for example being divided into three parts, I adopted these means without modification, deeming slight defects in coincidence not sufficiently important to be taken account of.

The period chosen for the research commenced with the year 1872, because although observations of the United States Weather Bureau date from 1871, when they were commenced by the Army Signal Service, the data for that year were insufficient. This consideration was paramount in preparing the work because, in first planning the work, it was not intended to include any stations but those for which uniform records were readily obtainable. It was also intended to include as many regions as possible in the investigation, but the circumstances mentioned in § 6 led to the omission of several regions which might have been included had the data been available. It was also believed that definitive results would be obtained by confining the discussion to those regions where the data were easily accessible and undoubted.

The regions and stations finally chosen were as follows :

1. *The United States East of the Rocky Mountains, Called U. S. I.* — In order to lessen the effect of accidental fluctuations at a single point several stations as widely separated as possible are preferable. Guided by the consideration that stations near the tropics were to be preferred, the four finally chosen for this region were Washington, Key West, Galveston and Saint Louis.

2. *The United States West of the Rocky Mountains, or U. S. II.* — The best station in this region was San Diego owing not only to its southern position, but to its compara-

tive steadiness of temperature. The peculiar climate of San Francisco seemed to render it inadvisable to adopt it as a station. The interior points of Salt Lake City and Phoenix, Arizona, were also selected and used as stations, although the observations at each point have suffered some interruption.

3. *The Argentine Republic.*—The main source for this region has been, as mentioned in Chapter II, the publications of the *Officina Meteorologica Argentina*. The number of stations that could be used was different in different years, and fell off to a single one in 1898.

4. *Samoa.* — The *Deutsche Uberseeische Meteorologische Beobachtungen* contain meteorological observations at a number of coast and island stations, but, for the most part, the observations were not pursued continuously through a sufficient period to be well adapted to the present work. The best station for our purpose proved to be Apia, where the record is nearly complete since 1890. The unpublished results for this station up to 1904 were courteously communicated by the director of the *Deutsche Seewarte* at Hamburg.

As no general principle is illustrated by the process of forming means and finding deviations from them by simple subtraction, the writer conceives that the purpose of the present work will be best subserved by omitting these merely routine details. If, as he earnestly hopes, some authority fully equipped with the necessary computing assistance shall in the interest of meteorology reconstruct the work in question, it can now be more thoroughly done than the author has succeeded in doing. Data continually accumulate from year to year and the results of the present work will, it is hoped, be found useful in any such reconstruction. As one of the special purposes now in view is to show the method of determining correlations, that purpose will be best subserved by excluding details not peculiar to the work itself. Some remarks on a few special points may however be made.

After the means were taken for the regions U. S. I, it was found that the accidental deviations at St. Louis were so much larger than at the other stations that the means would be more accordant if this station were omitted entirely. Its weight was therefore reduced to one third and new means taken.

After the definitive means had been formed, it was found that the fluctuations of temperature at Galveston, which were in general quite small, sometimes showed abnormally negative values. When this anomaly was specially noted, and the correctness of the record ascertained, it was too late to modify the work. The most plausible explanation which I can assign for these anomalous temperatures is that they are produced by the "northers" which are known to occasionally come down from the Rocky Mountain region into Texas, but which I did not suppose extended so far south as Galveston. The further examination of this point must be left to meteorologists.

§ 14. *Tabular Exhibit of Ten-day Departures During the Period 1872 to 1904.*

The original departures are shown in the following tables in the form which seemed best adapted to facilitate a critical examination and working out of the results. The means are the unweighted ones of the several regions, and are therefore the values of τ to be used in the formulæ of § 4.

The regions are: Eastern United States, Western United States, Argentina and Apia.

At the bottom of each annual column is given the algebraic sum of the departures, which will be useful in any test to which the work may be submitted. By dividing these terms by 36 we have annual deviations for the different regions, which should not differ much from those used in chapter III.

The comparison of the sum of the means with the mean of the sums may be used to test the accuracy of the computation.

Below each sum is given the sum of the squares of the 36 departures. These are used in the formulæ of § 4.

TABLE XI.

Simultaneous Departures of Temperature in Regions in ° C.

		1872				1873				1874				1875			
		U. S. I	U. S. II	Arg.	Mean	U. S. I	U. S. II	Arg.	Mean	U. S. I	U. S. II	Arg.	Mean	U. S. I	U. S. II	Arg.	Mean
Jan.	a	+1.2	−1.2	+0.3	+ 0.1	−0.6	+1.9	+0.4	+ 0.6	+ 2.3	+0.1	− 0.4	+0.7	− 2.6	− 0.1	+0.3	−0.8
	b	+0.4	−0.3	−1.4	− 0.4	−0.3	+2.1	+1.4	+ 1.1	− 1.0	+1.4	− 1.1	−0.2	− 2.4	+ 0.6	+0.5	−0.4
	c	−4.1	−1.6	0.0	− 1.9	−2.7	+0.5	+0.9	− 0.4	+ 2.7	−0.1	− 1.2	+0.5	+ 1.8	− 0.4	+0.7	+0.7
Feb.	a	−2.8	+0.4	+0.2	− 0.7	+0.8	+0.2	−2.1	− 0.4	− 1.2	−0.6	− 0.9	−0.9	− 2.9	− 0.2	+1.6	−0.5
	b	−1.6	+0.3	−0.2	− 0.5	+1.0	−2.0	−0.9	− 0.6	+ 2.2	−0.8	+ 0.4	+0.6	− 2.2	+ 0.4	+1.6	−0.1
	c	−0.2	−1.1	−3.5	− 1.6	−1.0	−0.4	0.0	− 0.5	+ 2.0	−2.2	+ 2.0	+0.6	+ 1.1	− 0.4	−1.1	−0.1
Mar.	a	−2.8	+0.7	+0.6	− 0.5	−1.4	+0.1	−1.8	− 1.0	+ 2.0	−2.5	3.2	−1.2	− 1.2	− 1.9	−0.2	−1.1
	b	−1.4	−0.2	−2.1	1.2	+1.3	+1.4	+0.3	+ 1.0	+ 2.1	−2.4	− 2.1	−0.8	0.0	− 2.9	−0.7	−1.2
	c	−0.7	+0.3	−3.2	1.2	0.0	+1.4	−1.7	− 0.1	+ 1.2	−1.3	− 0.6	−0.2	+ 0.6	− 1.2	−1.2	−0.6
April	a	+2.2	−2.0	+1.0	+ 0.4	+1.5	−0.8	−0.7	0.0	− 0.9	−0.6	− 0.5	−0.7	+ 0.6	− 2.8	+0.9	−0.4
	b	+1.6	−1.5	+3.5	+ 1.2	−1.6	+1.3	−1.0	− 0.4	− 1.0	−2.3	+ 0.6	−0.9	− 3.4	+ 1.7	−0.3	−0.7
	c	+1.7	−0.2	−2.6	− 0.4	−1.9	+0.9	−0.1	− 0.4	− 2.0	+0.8	− 1.5	+1.2	− 2.8	+ 1.4	−1.5	−1.0
May	a	+1.0	+0.8	+0.2	+ 0.7	−1.2	+1.1	+0.9	+ 0.3	− 1.2	−0.2	+ 1.3	0.0	− 0.3	+ 1.2	+1.8	+0.9
	b	+1.3	−2.5	−2.6	− 1.3	+0.2	−0.1	+0.6	+ 0.2	+ 0.3	+1.8	− 2.1	0.0	+ 0.7	+ 1.8	−1.1	0.0
	c	+1.3	+0.4	+0.6	+ 0.8	+1.9	+0.1	+1.5	+ 1.2	+ 0.9	+0.9	− 2.1	−0.1	+ 1.0	+ 0.8	+3.2	+1.7
June	a	+0.9	+0.3	−0.1	+ 0.4	+1.0	+0.1	+1.2	+ 0.8	+ 1.9	−0.1	+ 2.3	+1.4	− 0.4	+ 0.3	0.0	0.0
	b	+0.8	+1.0	+4.1	+ 2.0	+0.7	−0.4	−0.8	− 0.2	+ 0.2	+0.6	− 1.9	−0.4	− 0.3	+ 1.1	−2.6	−0.6
	c	+1.2	+2.9	−2.2	+ 0.6	+0.9	+0.2	+2.2	+ 1.1	+ 1.4	+0.2	+ 0.5	+0.7	+ 1.4	+ 1.2	−3.2	−0.2
July	a	+1.4	−0.1	−0.2	+ 0.4	+0.7	−0.2	−1.5	− 0.3	+ 0.4	+2.9	− 1.4	+0.6	+ 0.1	− 0.2	+1.5	+0.5
	b	+1.2	+0.5	+2.2	+ 1.3	+0.5	−0.9	−1.2	− 0.5	− 0.4	+1.3	− 0.2	+0.2	+ 0.2	+ 0.6	−1.9	−0.4
	c	+1.0	−0.2	−0.6	+ 0.1	+0.4	+0.9	−0.5	+ 0.3	− 0.1	+0.2	+ 0.3	+0.1	− 0.3	+ 0.3	+0.1	0.0
Aug.	a	+1.2	+0.2	+1.9	+ 1.1	+0.5	−0.3	−1.1	− 0.3	+ 0.1	−0.2	+ 4.4	+1.4	− 0.9	+ 2.8	+0.7	+0.9
	b	+2.0	−1.2	−0.4	+ 0.1	−0.1	+0.4	−0.6	− 0.1	+ 0.8	−0.2	− 1.0	0.0	− 0.3	+ 1.3	−1.9	−0.3
	c	+1.4	+1.8	0.0	+ 1.1	+1.0	+0.7	+1.6	+ 1.1	− 0.8	−0.2	+ 2.8	+0.6	− 1.1	− 0.6	+0.2	−0.5
Sept.	a	+1.3	+0.1	−0.1	+ 0.4	+0.4	−0.3	+2.8	+ 1.0	+ 0.4	−1.0	− 0.5	−0.4	+ 1.4	+ 1.6	−1.5	+0.5
	b	−0.1	+0.6	−0.1	+ 0.1	−0.9	+1.8	+1.2	+ 0.7	+ 0.8	−2.3	+ 0.6	−0.3	− 2.4	+ 1.3	+2.5	+0.5
	c	+1.7	−1.0	+0.6	+ 0.4	+1.0	+0.3	−0.5	+ 0.3	− 0.4	+0.9	− 0.6	0.0	− 2.6	+ 1.6	+1.4	+0.1
Oct.	a	+0.6	+0.2	+3.0	+ 1.3	−0.9	−0.8	−0.8	− 0.8	− 0.4	+1.8	+ 1.8	+1.1	− 0.4	+ 3.3	0.0	+1.0
	b	−1.4	−0.4	+2.4	+ 0.2	+0.3	−0.1	+0.5	+ 0.3	− 1.8	+1.4	− 3.4	−1.3	− 2.5	+ 4.6	+0.8	+1.0
	c	−0.1	−0.6	0.0	− 0.2	−2.8	−0.6	+1.1	− 0.8	+ 2.1	+0.4	− 2.1	+0.1	+ 1.1	+ 1.8	−1.2	+0.6
Nov.	a	−0.3	−0.4	−0.6	− 0.4	−0.3	0.0	+0.9	+ 0.2	+ 1.2	−0.7	− 2.8	−0.8	− 1.7	+ 0.5	+0.2	−0.3
	b	−3.8	+0.9	−1.3	− 1.4	−2.3	+2.7	+1.4	+ 0.6	+ 0.4	−0.4	+ 0.2	+0.1	+ 1.5	+ 1.1	+1.4	+1.3
	c	−1.5	−0.4	0.0	− 0.6	−0.7	+0.6	+0.8	+ 0.2	+ 0.1	+1.7	+ 0.3	+0.7	+ 0.5	+ 1.1	+0.2	+0.6
Dec.	a	−0.4	+0.7	0.0	+ 0.1	+2.6	−0.8	+0.7	+ 0.8	+ 0.9	+0.8	+ 1.2	+1.0	− 0.1	+ 1.8	+1.4	+1.0
	b	−1.9	−2.7	−0.5	1.7	+2.2	−1.7	−0.5	0.0	+ 0.6	−1.7	+ 0.8	−0.1	+ 0.6	− 0.1	−3.3	−0.9
	c	−4.9	+1.2	−0.4	1.4	−1.4	+0.3	+1.1	0.0	+ 2.2	−1.4	− 0.4	+0.1	+ 5.5	+ 1.2	+0.2	+2.3
Sum		−2.6	−4.5	−1.5	− 2.6	−1.2	+9.8	+5.7	+ 5.0	+17.9	−3.6	−10.5	+1.2	−14.1	+24.6	−0.5	+3.5
Σv²		120	44	97	41.42	62	38	48	12.89	70	64	102	18.15	114	96	81	24.01

TABLE XI.—Continued.

Simultaneous Departures of Temperature in Regions in °C.

		1876				1877				1878				1879			
		U.S. I	U.S. II	Arg.	Mean	U.S. I	U.S. II	Arg.	Mean	U.S. I	U.S. II	Arg.	Mean.	U.S. I	U.S. II	Arg.	Mean
Jan.	a	+4.8	+1.4	+0.5	+2.2	−4.7	+ 0.8	+ 1.1	− 0.9	− 3.8	−2.8	−0.1	−2.2	−5.2	− 1.2	−3.2	− 3.2
	b	+1.7	−1.0	−1.5	−0.3	+1.4	+ 0.5	+ 1.1	+ 1.0	+ 1.3	+0.7	−0.6	+0.5	−1.1	− 1.7	+0.2	− 0.9
	c	+4.2	−0.9	−1.7	+0.5	+1.0	− 0.8	+ 3.2	+ 1.1	+ 1.4	+2.4	−3.6	+0.1	+2.6	+ 1.8	−0.5	+ 1.3
Feb.	a	+0.4	+0.9	−0.8	+0.2	+2.7	+ 1.2	− 0.6	+ 1.1	0.0	+0.9	−1.5	−0.2	−0.9	− 1.1	+0.6	− 0.5
	b	+2.8	+1.4	−1.1	+1.0	−0.6	+ 1.2	− 0.3	+ 0.1	+ 0.3	+1.2	−0.5	+0.3	−1.4	+ 2.6	−0.5	+ 0.2
	c	+0.7	+0.1	+1.0	+0.6	−0.7	+ 2.7	− 0.4	+ 0.5	+ 2.0	−0.1	+0.2	+0.7	−1.6	+ 3.5	−1.6	+ 0.1
Mar.	a	+2.0	−1.2	+0.1	+0.3	−0.3	+ 1.8	+ 1.3	+ 0.9	+ 1.4	−0.4	+2.1	+1.0	+2.0	+ 2.4	+0.4	+ 1.6
	b	−0.3	−1.8	−0.5	−0.9	−1.2	+ 4.0	+ 2.9	+ 1.9	+ 2.5	+3.0	+0.6	+2.0	−0.6	+ 1.7	−0.1	+ 0.3
	c	−2.1	−1.2	+2.1	−0.4	−0.7	+ 3.0	+ 1.5	+ 1.3	+ 1.9	+0.2	−0.4	+0.6	+1.8	+ 3.4	−0.1	+ 1.7
April	a	−0.5	−0.9	+0.4	−0.3	−0.1	+ 0.9	− 0.7	0.0	+ 1.2	+1.8	−1.9	+0.4	−1.2	+ 1.2	−0.7	− 0.2
	b	0.0	+0.6	−0.2	+0.1	−0.4	+ 0.2	+ 1.7	+ 0.5	+ 2.0	−2.4	+0.1	−0.1	−0.8	+ 0.2	+0.2	− 0.1
	c	−0.1	+3.4	−0.2	+1.0	−0.6	− 1.3	+ 1.0	− 0.3	+ 2.0	−0.8	+0.8	+0.7	+1.0	+ 1.6	+1.0	+ 1.2
May	a	−0.3	+1.3	+2.3	+1.1	−3.3	− 0.1	− 0.6	− 1.3	+ 0.9	+1.1	+0.1	+0.7	−0.4	+ 2.2	+0.9	+ 0.9
	b	+0.3	−0.8	+1.7	+0.4	+0.3	− 2.2	− 1.2	− 1.0	− 0.7	0.0	−2.5	−1.1	+1.7	+ 1.0	+1.2	+ 1.3
	c	+0.2	+0.6	+0.5	+0.4	+0.1	− 0.7	+ 0.1	− 0.2	+ 0.4	−0.9	+2.9	+0.8	+0.2	− 1.7	+0.1	− 0.5
June	a	+0.6	−0.2	+1.1	+0.5	+0.3	+ 0.3	+ 1.6	+ 0.7	+ 0.1	+1.0	−0.8	+0.1	−0.3	+ 1.8	−2.2	− 0.2
	b	+0.4	+2.8	−1.8	+0.5	+0.3	+ 2.0	− 1.1	+ 0.4	− 0.3	+1.2	+2.5	+1.1	0.0	− 1.4	+2.0	+ 0.2
	c	+1.0	+1.3	+1.0	+1.1	+0.4	− 0.5	+ 1.9	+ 0.6	+ 0.4	+1.4	−3.3	−0.5	−0.3	+ 0.7	+0.1	+ 0.2
July	a	+1.4	+1.6	+0.6	+1.2	+0.8	+ 1.5	+ 2.4	+ 1.6	+ 0.9	+1.7	−2.5	0.0	+0.7	− 0.1	+1.4	+ 0.7
	b	+1.3	+0.7	+4.3	+2.1	−0.1	+ 2.8	+ 6.4	+ 3.0	+ 2.0	+1.0	+1.3	+1.4	+0.7	+ 0.6	+1.7	+ 1.0
	c	−0.8	+0.6	+3.1	+1.0	+0.7	+ 0.2	− 2.5	− 0.5	+ 0.8	+0.8	+0.8	+0.8	+0.3	+ 1.2	+0.7	+ 0.7
Aug.	a	−0.5	−0.1	−0.1	−0.2	+0.5	+ 0.6	− 3.0	− 0.6	+ 1.0	+1.6	+3.0	+1.9	+0.6	+ 0.6	−1.8	− 0.2
	b	+0.5	−0.1	−2.4	−0.7	+0.3	+ 1.6	+ 1.8	+ 1.2	+ 0.7	+1.7	−2.6	−0.1	−1.2	+ 1.4	+1.2	+ 0.5
	c	+0.4	−1.1	+0.7	0.0	+1.4	+ 0.5	+ 0.4	+ 0.8	+ 0.2	+0.5	−1.5	−0.3	−0.1	− 0.1	+2.8	+ 0.9
Sept.	a	+0.6	−0.7	+0.2	0.0	+0.1	+ 1.0	+ 1.1	+ 0.7	+ 0.9	−1.5	+1.3	+0.2	−0.2	+ 1.6	−1.7	− 0.1
	b	−0.9	−0.3	+0.7	−0.2	+0.7	− 0.4	+ 0.5	+ 0.3	− 0.9	−0.2	−0.1	−0.4	−1.6	+ 2.1	−0.1	+ 0.1
	c	+0.3	+1.2	+2.6	+1.2	+0.7	+ 0.3	− 2.0	− 0.3	+ 0.2	−1.2	−0.4	−0.5	−0.4	+ 1.0	+1.1	+ 0.6
Oct.	a	−2.1	+2.9	+2.1	+1.0	−0.2	+ 0.4	+ 1.0	+ 0.4	+ 0.6	−0.1	0.0	+0.2	+3.3	− 0.3	+2.4	+ 1.8
	b	−1.6	+0.1	−2.4	−1.3	+1.2	− 0.6	+ 0.3	+ 0.3	+ 1.7	−0.6	−0.3	+0.3	+2.7	− 0.6	−1.1	+ 0.3
	c	+0.8	0.0	−1.2	−0.1	+0.8	− 1.0	+ 1.5	+ 0.4	− 1.4	−1.3	−1.0	−1.2	−1.3	+ 1.3	−2.0	− 0.7
Nov.	a	+0.1	+0.2	−2.1	−0.6	−1.2	− 0.8	− 1.1	− 1.0	− 1.0	+0.5	+0.2	−0.1	−0.4	− 0.6	+1.0	0.0
	b	−0.7	+0.3	−2.8	−1.1	+0.5	+ 0.4	+ 2.0	+ 1.0	+ 0.7	+0.1	+2.1	+1.0	+3.0	− 1.4	+1.7	+ 1.1
	c	−1.5	+0.5	−1.1	−0.7	−0.3	− 1.3	− 0.2	− 0.6	+ 0.3	−1.0	−0.3	−0.3	0.0	− 2.0	−0.2	− 0.7
Dec.	a	−5.9	+1.0	−1.2	−2.0	−2.2	− 2.4	− 0.6	− 1.7	− 0.1	−0.8	−2.2	−1.0	+3.6	− 0.7	−1.1	+ 0.6
	b	−1.6	−2.9	−2.7	−2.4	+3.3	+ 2.2	+ 1.8	+ 2.4	− 1.9	−4.4	−0.3	−2.2	+0.7	+ 1.3	+0.0	+ 0.7
	c	−3.7	−0.3	−0.6	−1.5	+2.3	− 0.4	+ 0.7	+ 0.4	− 4.5	−3.0	−1.5	−3.0	+1.1	+ 3.2	+1.0	+ 0.4
Sum		+1.3	+9.4	+0.6	+3.7	+3.1	+17.6	+21.6	+14.2	+13.2	+1.2	−9.9	+1.6	+8.0	+19.1	+4.8	+10.3
ΣΔ²		124	58	96	37.61	76	81	113	41.18	84	89	97	40.83	101	98	63	32.92

TABLE XI. — *Continued.*

Simultaneous Departures of Temperature in Regions in °C.

		1880				1881				1882				1883			
		U. S. I	U. S. II	Arg.	Mean	U. S. I	U. S. II	Arg.	Mean	U. S. I	U. S. II	Arg.	Mean	U. S. I	U. S. II	Arg.	Mean
Jan.	a	+ 5.8	+ 0.1	−0.6	+1.8	− 3.6	−2.8	−0.7	−2.4	+ 1.0	− 1.2	−0.7	−0.3	−1.1	−0.4	+3.6	+0.7
	b	+ 4.3	+ 1.6	−1.6	+1.4	− 1.3	+0.4	−2.6	−1.2	+ 2.6	− 3.9	+1.1	−0.1	−1.3	−2.9	−1.7	−2.0
	c	+ 4.0	− 1.6	0.0	+0.8	− 1.8	−0.7	−0.4	−1.0	+ 0.9	− 2.8	+0.4	−0.5	−0.7	−0.9	−0.9	−0.8
Feb.	a	+ 0.9	− 4.2	+1.1	−1.3	− 1.5	+2.9	0.0	+0.5	+ 2.0	− 2.8	−2.3	−1.0	−0.2	−3.9	−0.1	−1.4
	b	+ 2.3	− 3.4	−1.1	−0.7	− 1.0	−1.4	+1.4	−0.3	+ 4.6	− 2.0	+0.5	+1.0	+2.6	−1.7	−0.7	+0.1
	c	+ 3.7	− 1.7	−1.3	+0.2	− 0.7	+2.7	+1.1	+1.0	+ 1.6	− 1.3	+0.4	+0.2	+0.3	+0.2	+0.7	+0.4
Mar.	a	+ 3.4	− 1.6	−2.2	−0.1	− 1.0	−1.1	+0.1	−0.5	+ 2.6	− 2.8	−1.1	−0.4	−0.7	+2.0	+1.6	+1.0
	b	− 1.1	− 5.0	0.0	−2.0	+ 0.7	−3.8	+2.7	−0.1	+ 1.9	− 1.0	−0.5	+0.1	−0.1	+1.9	−0.8	+0.3
	c	+ 0.2	− 1.9	+0.3	−0.5	− 2.5	+1.8	+0.9	+0.1	+ 0.9	+ 1.4	−0.2	+0.7	−2.0	+1.8	+2.0	+0.6
April	a	+ 1.3	+ 0.8	+0.7	+0.9	− 3.9	+1.9	+0.2	−0.6	+ 2.9	+ 0.4	−1.3	+0.7	+1.2	−1.0	+0.6	+0.3
	b	+ 1.2	− 1.8	+0.6	0.0	− 1.2	+0.7	−0.6	−0.4	− 1.2	− 2.6	−1.9	−1.9	+0.9	−2.5	−0.2	−0.6
	c	+ 0.6	− 1.4	−1.5	−0.8	+ 1.8	+2.7	+1.0	+1.8	− 0.2	− 0.7	−1.3	−0.7	−1.1	−2.0	−2.6	−1.9
May	a	+ 1.9	+ 0.3	+0.2	+0.8	+ 0.7	+2.1	−0.3	+0.8	+ 0.2	+ 0.5	−0.8	0.0	+0.9	−1.6	+1.9	+0.4
	b	+ 0.9	− 0.4	+0.3	+0.3	+ 1.6	+0.2	+1.3	+1.0	− 2.7	+ 0.8	+0.5	−0.5	−0.4	−1.8	+1.3	−0.3
	c	+ 1.8	+ 0.5	+1.0	+1.1	+ 1.3	+1.3	+0.8	+1.1	− 1.2	+ 0.2	+2.0	+0.3	−1.3	+0.9	−1.0	−0.5
June	a	+ 0.4	+ 0.7	+3.4	+1.5	+ 0.4	+1.0	+0.1	+0.5	− 0.6	+ 1.7	−2.0	−0.3	+0.9	+1.0	+2.1	+1.3
	b	+ 0.5	+ 1.0	+1.6	+1.1	+ 1.4	+0.7	+0.2	+0.8	+ 0.7	− 1.3	0.0	−0.2	+0.8	+0.6	+2.4	+1.3
	c	+ 0.6	+ 1.4	+2.0	+1.3	+ 0.7	+1.0	−1.1	+0.2	+ 1.6	+ 0.1	+1.1	+0.9	+0.4	+1.7	−0.7	+0.5
July	a	+ 0.5	+ 0.6	+0.3	+0.5	+ 0.7	+0.1	+0.1	+0.3	− 0.5	+ 1.3	+1.4	+0.7	+0.6	+1.2	+2.1	+1.3
	b	+ 1.0	− 1.4	+2.5	+0.7	+ 1.0	−0.2	−1.6	−0.3	− 0.4	− 0.1	−0.2	−0.2	0.0	+0.4	+2.0	+0.8
	c	+ 0.6	− 0.9	−0.6	−0.7	+ 0.2	−1.0	−1.2	−0.7	− 0.1	− 0.5	−2.7	−1.1	+0.1	−0.6	−4.2	−1.6
Aug.	a	− 1.3	− 1.3	+0.4	−0.7	+ 1.2	0.0	−3.8	−0.9	− 0.7	+ 1.8	+1.5	+0.9	−0.5	−0.6	−0.8	−0.6
	b	− 0.2	+ 0.1	+1.8	+0.6	+ 0.2	−1.0	−0.3	−0.4	− 0.2	+ 0.2	+0.7	+0.2	−0.1	−0.1	−0.6	−0.3
	c	+ 1.4	− 1.7	+2.3	+0.7	+ 1.7	−0.9	+3.6	+1.5	− 0.1	+ 0.1	+0.1	0.0	+0.1	+2.0	−0.3	+0.6
Sept.	a	− 0.2	+ 0.1	−0.5	−0.2	+ 2.8	−2.3	−1.0	−0.2	+ 0.2	+ 1.7	−1.9	0.0	−0.9	+2.8	−2.5	−0.2
	b	− 0.1	− 0.8	−5.3	−2.2	+ 0.4	+0.4	+0.3	+0.4	+ 0.8	− 1.9	−0.3	−0.5	+0.3	+0.3	+0.2	+0.3
	c	+ 0.2	− 0.5	+1.0	+0.2	+ 2.1	−2.6	−0.8	−0.4	− 0.9	+ 0.2	+1.8	+0.4	−0.4	+2.6	+0.8	+1.0
Oct.	a	− 0.6	− 0.1	−2.6	−1.1	+ 2.7	+1.4	+1.2	+1.8	+ 1.5	− 2.7	+2.2	+0.3	+0.8	−2.0	+2.4	+0.4
	b	− 0.1	− 2.6	−2.1	−1.6	+ 2.3	−2.9	+1.6	+0.3	+ 1.6	− 1.6	+1.2	+0.4	+1.3	−2.7	+1.3	0.0
	c	− 1.5	+ 0.8	−1.7	−0.8	+ 1.3	−1.9	−1.1	−0.6	+ 1.6	+ 1.0	+2.8	+1.8	+1.1	−1.7	−2.6	−1.1
Nov.	a	− 0.2	− 0.8	+1.7	−0.2	+ 1.4	−2.3	0.0	−0.3	+ 1.9	+ 0.6	+0.8	+1.1	+1.6	−0.1	+2.1	+1.2
	b	− 3.0	− 5.0	−1.2	−3.1	+ 1.7	−2.7	+1.2	+0.1	− 0.5	− 3.7	−0.3	−1.5	−2.1	−0.2	−0.7	−1.0
	c	− 3.2	− 4.3	−0.8	−2.7	− 1.0	−2.3	−0.2	−1.2	− 2.0	− 0.1	+0.2	−0.6	+1.4	−0.8	−0.4	+0.1
Dec.	a	− 0.7	− 2.3	+2.7	−0.1	+ 1.8	+0.9	+2.0	+1.6	− 1.8	+ 2.0	−0.3	0.0	+0.7	−0.6	+1.6	+0.6
	b	+ 0.3	+ 0.4	+0.4	+0.4	+ 2.3	−0.3	+1.8	+1.3	− 0.7	+ 2.7	−1.3	+0.2	−0.8	+1.4	+1.5	+0.7
	c	− 5.8	+ 1.6	−0.3	−1.5	+ 0.2	+0.6	+1.3	+0.7	+ 0.3	− 1.0	−1.5	−0.7	+1.6	+0.6	−0.8	+0.5
Sum		+16.6	−34.7	+0.4	−5.9	+13.3	−7.6	+7.8	+4.3	+17.6	−17.3	−1.9	−0.6	+3.9	−6.7	+8.5	+2.1
$\Sigma\Delta^2$		172	149	109	51.52	104	113	70	31.69	91	114	62	19.76	40	99	108	28.65

TABLE XI.—*Continued.*

Simultaneous Departures of Temperature in Regions in °C.

		1884				1885				1886				1887			
		U. S. I	U. S. II	Arg.	Mean	U. S. I	U. S. II	Arg.	Mean	U. S. I	U. S. II	Arg.	Mean	U. S. I	U. S. II	Arg.	Mean
Jan.	a	− 2.3	+1.1	−0.5	−0.6	+ 0.9	+ 0.3	+0.1	+0.4	− 2.3	−3.9	+2.0	−1.4	−4.4	+ 1.7	− 0.8	−1.2
	b	− 1.0	−1.2	+2.7	+0.2	+ 0.2	+ 0.2	+0.4	+0.3	− 3.2	+0.7	+0.9	−0.5	− 0.3	+ 2.0	+ 1.6	+1.1
	c	− 1.5	+1.0	+2.5	+0.7	− 2.2	− 1.0	+2.3	−0.3	− 1.8	+4.9	+0.5	+1.2	+ 3.0	+ 1.2	+ 1.4	+1.9
Feb.	a	+ 4.0	+1.0	−0.9	+1.4	+ 0.5	+ 2.0	+0.7	+1.1	− 4.2	+4.3	−0.3	−0.1	+ 2.8	+ 2.3	− 0.4	+1.6
	b	+ 2.2	−3.8	+1.6	0.0	− 2.3	+ 0.4	−1.6	−1.2	− 0.4	+2.9	−2.3	+0.1	+ 3.4	− 0.3	− 0.2	+1.0
	c	− 1.3	+2.2	+0.5	+0.5	− 3.5	+ 1.2	−0.9	−1.1	− 1.4	+1.7	+3.2	+1.2	+ 0.6	− 1.4	− 0.8	−0.5
Mar.	a	− 1.5	+3.1	+1.4	+1.0	− 1.1	+ 1.9	−0.5	+0.1	− 2.6	−2.0	+1.6	−1.0	+ 1.9	+ 2.5	− 0.1	+1.4
	b	+ 1.2	−0.3	+2.3	+1.1	− 3.9	+ 2.2	0.0	−0.6	+ 0.8	−1.8	+1.3	+0.1	+ 0.9	+ 2.0	+ 0.9	+1.3
	c	+ 2.4	−1.9	+2.1	+0.9	− 1.9	+ 1.7	0.0	−0.1	− 0.9	−1.1	0.0	−0.7	+ 2.4	+ 1.1	+ 1.4	0.0
April	a	− 1.3	+0.6	−2.2	−1.0	+ 1.0	+ 3.6	−0.4	+1.4	− 2.6	−0.3	+1.4	−0.5	− 0.7	+ 1.4	− 0.5	+0.1
	b	− 1.3	−0.8	−1.0	−1.0	− 0.9	+ 1.9	+4.2	+1.7	+ 0.4	−1.3	0.0	−0.3	0.0	− 1.8	− 1.0	−0.9
	c	− 1.4	−1.4	+2.0	−0.3	+ 1.4	+ 0.1	−0.4	+0.4	+ 0.9	−0.9	−1.5	−0.5	− 0.4	− 0.3	− 1.0	−0.6
May	a	+ 0.6	−0.4	−3.6	−1.1	− 0.5	+ 2.1	0.0	+0.5	0.0	−0.1	−1.4	−0.5	+ 1.6	+ 1.0	− 2.3	+0.1
	b	+ 0.4	+1.1	−1.2	+0.1	− 0.4	− 1.0	−1.6	−1.0	− 0.9	0.0	−0.6	−0.5	+ 0.4	0.0	+ 0.3	+0.2
	c	− 0.2	+0.1	+0.4	+0.1	+ 0.4	− 0.4	−1.0	−0.3	− 0.4	+2.8	+2.6	+1.7	+ 0.1	+ 1.9	+ 1.8	+1.3
June	a	+ 0.4	+0.7	+1.8	+1.0	+ 0.2	− 1.1	+0.3	−0.2	− 0.3	+1.1	−0.6	+0.1	− 0.6	+ 0.2	+ 3.6	+1.1
	b	− 0.6	−0.2	−2.3	−1.0	+ 0.7	− 1.2	−4.4	−1.6	+ 0.3	+1.1	−0.9	+0.2	+ 0.5	+ 0.8	+ 3.8	+1.4
	c	+ 0.1	+1.2	−4.6	−1.1	− 0.1	− 0.2	+0.2	0.0	− 0.7	+1.3	−0.5	0.0	− 1.7	+ 1.0	+ 1.1	+0.1
July	a	+ 0.2	+0.9	0.0	+0.4	+ 0.3	+ 0.9	+0.9	+0.7	− 0.3	+1.0	−2.6	−0.6	− 0.1	+ 0.8	− 1.5	−0.3
	b	− 0.4	−0.4	−0.8	−0.5	+ 0.8	+ 0.4	−3.5	−0.8	− 0.9	+2.2	−1.6	−0.1	+ 0.8	− 1.1	− 0.8	−0.4
	c	+ 0.4	−0.8	−1.4	−0.1	+ 1.3	− 0.1	−3.0	−0.6	+ 0.4	−0.6	+1.9	+0.6	+ 0.7	− 0.3	+ 1.9	+0.8
Aug.	a	− 0.7	−0.3	+3.6	+0.9	+ 0.2	− 0.4	−3.5	−1.2	− 0.4	−0.2	−0.4	−0.3	+ 0.2	− 1.4	+ 4.9	+1.2
	b	+ 0.1	−0.8	+2.4	+0.6	− 0.2	+ 2.6	−0.7	+0.6	+ 0.2	+0.9	−0.7	+0.1	+ 0.3	0.0	+ 4.7	+1.7
	c	+ 0.9	0.0	+3.0	+1.3	+ 0.4	− 0.5	+1.8	+0.6	+ 0.6	+1.7	+1.0	+1.1	− 1.0	− 1.2	− 0.4	−0.9
Sept.	a	+ 1.3	−3.7	+1.0	−0.5	− 0.3	− 0.6	+0.5	−0.1	0.0	+1.7	+2.4	+1.4	− 0.3	− 1.4	− 1.9	−1.2
	b	+ 0.5	−1.8	+0.3	−0.3	+ 0.7	+ 0.3	+2.8	+1.3	+ 1.0	+0.8	−0.7	+0.4	− 0.1	+ 0.2	− 1.4	−0.4
	c	+ 2.4	−0.6	−1.2	+0.2	−.0.1	+ 1.4	−0.6	+0.2	+ 0.8	−0.4	−3.2	−0.9	− 1.6	+ 1.6	+ 1.6	+0.5
Oct.	d	+ 2.9	−3.1	−0.1	−0.1	− 1.1	+ 2.1	+2.2	+1.1	− 1.0	+0.2	−1.1	−0.6	+ 1.3	+ 0.2	+ 1.0	+0.8
	b	+ 1.5	+1.8	+0.4	+1.2	− 0.1	− 0.1	−0.3	−0.2	+ 1.6	−2.0	+0.8	+0.1	− 1.2	+ 0.2	− 0.7	−0.6
	c	+ 1.6	−0.1	−0.8	+0.2	− 1.4	+ 0.8	+0.2	−0.1	− 0.2	−1.0	−1.1	−0.8	− 2.3	+ 0.9	− 1.7	−1.0
Nov.	a	− 0.5	+0.3	−0.1	−0.1	+ 1.0	+ 1.0	+3.9	+2.0	+ 0.6	−2.3	+0.9	−0.3	− 0.2	+ 2.2	− 0.6	+0.5
	b	+ 0.6	+0.3	+0.7	+0.5	+ 0.8	+ 1.3	−0.6	+0.5	− 0.5	−4.6	−0.3	−1.8	+ 0.2	+ 1.3	+ 0.8	+0.8
		+ 0.3	+0.6	0.0	+0.3	− 1.4	+ 1.1	+1.9	+0.5	+ 0.4	−2.7	−0.1	−0.8	+ 0.2	− 0.2	− 0.8	−0.3
Dec.	a	+ 2.3	−1.6	+0.4	+0.4	− 1.7	− 0.3	+0.3	−0.6	− 4.0	−0.7	+1.0	−1.2	+ 0.9	− 0.4	− 0.2	+0.1
	b	− 1.4	+2.6	+0.6	+0.6	− 1.2	− 0.8	−0.8	−0.9	− 0.9	+0.2	+2.4	+0.9	− 0.1	− 1.6	+ 1.2	−0.2
	c	+ 0.4	+1.4	−0.1	+0.6	+ 1.5	+ 4.7	+0.2	+2.1	− 0.1	+2.7	+2.1	+1.6	− 2.2	− 1.4	− 1.1	−1.6
Sum		+11.4	−3.0	+8.9	+6.5	−11.9	+26.5	−0.9	+4.6	−22.1	+7.3	+5.9	−2.6	− 0.8	+13.7	+13.8	+8.9
ΣΔ²		76	87	115	19.45	64	85	120	29.98	80	154	85	25.40	83	58	119	31.94

TABLE XI. — *Continued.*

Simultaneous Departures of Temperature in Regions in ° C.

		1888				1889				1890					1891				
		U.S. I	U.S. II	Arg.	Mn.	U.S. I	U.S. II	Arg.	Mn.	U.S. I	U.S. II	Arg.	Samoa	Mn.	U.S. I	U.S. II	Arg.	Samoa	Mn.
Jan.	a	+ 1.4	− 3.5	+ 0.5	−0.5	+0.9	− 1.6	+1.4	+0.2	+ 5.9	− 2.4	− 0.6	....	+1.0	−0.1	+0.6	+1.1	+0.6	+0.6
	b	− 2.9	− 5.9	+ 3.1	−1.9	+2.4	+ 1.3	−0.1	+1.2	+ 4.4	− 3.3	− 2.6	....	−0.5	−1.3	−0.3	+0.4	+0.8	−0.1
	c	− 1.9	+ 2.8	+ 2.5	+1.1	+0.1	− 2.9	−0.8	−1.2	+ 3.4	− 0.1	+ 0.6	−0.3	+0.9	+1.9	+1.7	−0.6	−0.2	+0.7
Feb.	a	+ 1.1	+ 1.6	+ 0.2	+1.0	−1.2	− 0.9	−1.0	−1.0	+ 4.0	+ 5.0	+ 1.0	−0.1	+2.5	+2.2	−2.6	+1.1	0.0	+0.2
	b	+ 0.1	+ 2.4	+ 0.5	+1.0	−0.7	− 1.4	−0.5	−0.9	+ 2.2	− 0.4	+ 0.4	+0.2	+0.6	+2.6	−0.6	+1.1	0.0	+0.8
	c	+ 0.3	+ 1.7	+ 1.9	+1.3	−2.9	+ 2.7	−0.2	−0.1	+ 1.9	− 4.2	− 2.1	+0.0	−1.1	+0.2	+0.4	+1.9	+0.3	+0.7
Mar.	a	− 1.2	− 1.7	0.0	−1.0	−1.6	+ 3.4	+1.1	+1.0	− 2.9	− 0.4	− 1.0	+0.5	−0.9	−1.1	−1.7	+3.8	−0.1	+0.2
	b	− 2.2	+ 1.0	+ 1.5	+0.1	+0.3	+ 1.8	+0.4	+0.8	+ 0.1	+ 0.2	− 1.3	+0.3	−0.2	−1.4	+1.1	−1.1	+0.5	−0.2
	c	− 1.7	− 0.3	+ 0.6	−0.5	−0.5	+ 2.4	+1.1	+1.0	+ 2.1	− 0.2	− 1.9	+0.6	+0.1	−0.8	−1.9	+0.8	+0.3	−0.4
April	a	+ 2.2	+ 0.8	+ 0.7	+1.2	+0.2	+ 3.0	+1.8	+1.7	+ 0.9	+ 0.4	− 1.1	−0.7	−0.1	−3.4	−0.7	−1.9	+0.3	−1.4
	b	− 0.2	+ 4.4	+ 1.6	+1.9	+0.4	+ 0.1	−3.3	−0.9	− 0.1	− 0.2	+ 1.5	−0.8	+0.1	+1.6	−0.7	+1.2	−0.1	+0.5
	c	− 0.8	+ 2.2	+ 1.2	+0.9	−0.3	+ 3.1	−0.9	+0.6	− 0.6	+ 0.8	+ 3.0	+0.2	+0.8	+0.8	+1.4	−0.4	−0.1	+0.4
May	a	+ 0.1	+ 0.1	+ 2.2	+0.8	−0.7	− 0.6	+0.9	−0.1	− 0.3	+ 1.2	+ 1.4	−0.1	+0.6	−0.9	+2.1	−1.4	−0.2	−0.1
	b	− 1.6	+ 1.4	− 2.1	−0.8	+0.9	− 0.7	−1.9	−0.6	− 0.4	+ 0.4	− 2.0	+0.4	−0.4	−1.6	+0.4	−1.8	+0.1	−0.7
	c	− 0.5	− 0.3	− 1.0	−0.6	−0.3	+ 1.9	+2.3	+1.3	− 0.4	+ 1.1	− 1.5	−0.1	−0.2	−1.5	−0.7	+0.8	+0.3	−0.3
June	a	− 1.0	+ 0.4	− 2.7	−1.1	−1.6	+ 0.7	+1.5	+0.2	− 0.6	+ 0.7	0.0	+0.6	+0.2	−1.0	−1.1	0.0	+0.7	−0.4
	b	+ 0.1	+ 1.8	− 2.7	−0.3	−0.4	+ 1.2	−2.6	−0.6	− 0.3	− 0.9	− 4.0	−0.2	−1.2	+0.3	−1.3	+1.2	+0.2	+0.1
	c	+ 0.1	+ 1.0	− 2.2	−0.4	−1.6	+ 1.0	+0.4	−0.1	+ 0.7	− 1.2	− 2.6	0.0	−0.8	+0.1	+1.8	+0.8	−0.7	+0.5
July	a	− 0.7	+ 1.9	+ 4.1	+1.8	−0.3	+ 0.3	−2.6	−0.9	+ 0.1	+ 1.3	− 0.9	+0.4	+0.2	−2.0	−0.1	+1.5	0.0	−0.2
	b	− 1.0	− 0.2	+ 1.7	+0.2	0.0	+ 1.1	−0.9	+0.1	− 0.5	+ 0.4	+ 0.5	−1.3	−0.2	−0.3	−0.9	−1.2	−0.9	0.8
	c	− 0.7	+ 0.6	+ 0.8	+0.2	−0.6	+ 1.7	+1.4	+0.8	− 1.4	+ 1.7	+ 2.5	−0.6	+0.6	−0.7	+0.9	−1.6	−0.7	−0.5
Aug.	a	+ 0.8	− 1.0	+ 4.6	+1.5	−1.3	+ 1.1	+0.6	+0.1	− 0.1	− 0.3	− 1.6	−0.8	−0.7	−0.2	−0.7	−2.0	−0.2	−0.8
	b	− 0.2	− 0.5	− 1.5	−0.7	−1.3	+ 1.5	−4.8	−1.5	− 0.7	− 0.1	− 1.7	+0.5	−0.5	0.0	+1.2	−0.3	−0.8	0.0
	c	− 0.9	+ 1.2	+ 0.7	+0.3	−0.5	+ 1.4	−1.2	−0.1	− 0.8	+ 0.2	− 2.3	−0.2	−0.9	−1.4	+2.2	+1.5	−0.1	+0.6
Sept.	a	− 1.3	+ 1.7	+ 4.1	+1.5	−0.1	− 0.1	−1.9	−0.7	− 0.1	− 0.8	− 1.5	+0.9	−0.4	−1.3	+3.7	−3.8	+0.1	−0.3
	b	− 0.2	+ 2.0	+ 1.2	+1.0	−0.4	+ 0.6	+1.3	+0.1	− 0.5	+ 0.8	− 2.4	+0.1	−0.5	+1.0	+0.6	−0.2	+0.6	+0.5
	c	− 1.9	+ 2.3	− 2.3	−0.6	−2.1	+ 0.4	−0.4	−0.7	− 2.0	+ 2.5	+ 0.9	+0.8	+0.1	+1.3	−0.9	+3.1	−0.6	+0.7
Oct.	a	− 1.9	+ 0.6	− 0.2	−0.5	−1.9	+ 2.3	+2.4	+0.9	+ 1.0	− 2.2	+ 1.2	+0.2	0.0	−0.9	−1.8	+2.6	+0.5	+0.1
	b	− 0.8	+ 0.9	+ 0.7	+0.3	−1.3	+ 0.1	+1.4	+0.1	+ 1.2	− 2.0	+ 1.6	0.0	+0.2	−1.8	−0.1	−0.9	−0.1	−0.7
	c	− 0.1	+ 0.3	+ 0.3	+0.2	−0.5	+ 1.2	−0.4	+0.1	− 1.8	+ 2.1	− 1.2	+0.5	−0.1	−0.6	+3.0	+1.2	+1.2	+1.2
Nov.	a	+ 2.7	− 1.4	+ 1.9	+1.1	−0.7	− 0.7	−1.1	−0.8	+ 0.5	+ 0.8	+ 1.6	+0.8	+0.9	−0.4	+1.9	−2.8	+0.6	−0.2
	b	+ 0.2	− 0.4	− 2.4	−0.9	−0.1	0.0	+0.4	+0.1	+ 1.9	0.0	− 2.3	−0.8	−0.3	−1.0	−0.7	+2.2	+0.6	+0.3
	c	− 2.9	+ 1.2	+ 0.9	−0.3	+0.4	+ 2.4	+1.4	+1.4	+ 0.8	+ 3.9	+ 3.5	+0.2	+2.1	−0.3	+2.3	+0.3	−0.8	+0.4
Dec.	a	− 1.1	+ 0.8	+ 0.5	+0.1	+1.8	+ 3.0	−1.2	+1.2	− 1.0	+ 1.2	+ 3.0	+0.1	+0.8	+1.1	−1.6	−1.0	+0.2	−0.3
	b	− 0.5	+ 1.9	+ 1.6	+1.0	+3.0	+ 3.4	−1.4	+1.7	+ 0.6	+ 3.3	+ 2.6	+0.5	+1.4	−0.2	−0.4	−1.0	−0.4	−0.3
	c	− 0.1	− 0.7	+ 0.7	0.0	+5.2	+ 0.7	−0.6	+1.8	− 0.1	+ 3.6	+ 0.8	+0.5	+1.2	+3.1	−3.7	−1.8	+0.7	−0.4
Sum		−19.3	+21.1	+23.2	+8.4	−7.2	+33.7	−8.0	+6.2	+16.5	+12.5	−10.3	+2.3	+5.3	−6.5	+2.6	+2.8	+2.6	+0.4
ΣΔ²		64	131	136	32.24	80	116	93	29.34	126	127	132	7	27.35	69	96	100	5	11.09

TABLE XI.—*Continued.*

Simultaneous Departures of Temperature in Regions in ° C.

		1892					1893					1894					1895				
		U.S. I	U.S. II	Arg.	Samoa	Mean	U.S. I	U.S. II	Arg.	Samoa	Mean	U.S. I	U.S. II	Arg.	Samoa	Mn.	U.S. I	U.S. II	Arg.	Samoa	Mn.
Jan.	a	− 1.3	+0.1	+ 0.6	+0.1	− 0.1	−2.7	+ 2.2	+ 0.6	− 0.8	− 0.2	+2.7	− 3.5	+1.0	+0.2	+0.1	− 0.8	+ 1.3	− 1.8	−0.4	−0.4
	b	− 3.0	−1.6	+ 1.4	−0.8	− 1.0	−5.8	0.2	− 1.2	− 1.1	− 2.1	+3.3	− 0.9	−2.2	−0.1	0.0	+ 0.4	+ 2.6	− 3.4	−0.2	−0.2
	c	− 0.7	+0.2	− 2.0	+0.1	− 0.6	−0.1	+ 0.3	+ 0.5	− 0.2	+ 0.1	−0.4	+ 1.2	+2.2	−0.7	+0.6	− 0.7	− 3.0	− 0.3	+0.2	−1.0
Feb.	a	+ 1.8	−1.0	+ 2.8	+0.6	+ 1.0	+0.5	− 1.0	− 2.1	− 0.2	− 0.7	+1.5	− 2.3	+0.7	−0.6	−0.2	− 5.8	− 0.9	+ 1.1	−0.9	−1.6
	b	+ 0.5	−0.3	+ 1.7	+0.4	+ 0.6	+0.7	− 1.0	− 1.5	− 0.8	− 0.6	−0.1	− 3.6	+0.7	−0.5	−0.9	− 7.5	− 2.3	+ 1.2	−1.0	−3.0
	c	+ 0.2	+1.8	+ 0.2	+0.2	+ 0.6	−0.4	− 1.8	− 1.6	− 0.6	− 1.1	−2.8	− 3.3	−0.2	−0.1	−1.6	− 1.1	+ 1.8	+ 1.9	−1.2	+0.4
Mar.	a	− 0.9	+1.9	− 0.8	+0.7	+ 0.2	+0.3	− 1.6	− 0.2	+ 0.6	− 0.2	+3.3	− 2.4	−1.8	−1.1	−0.5	− 0.6	+ 0.4	+ 1.4	−0.8	+0.1
	b	− 4.7	+1.7	− 2.0	+0.5	− 1.1	−1.0	− 2.1	− 0.6	− 0.8	− 1.1	+3.2	− 0.8	−2.1	+0.2	+0.1	− 1.3	− 2.6	+ 0.9	−0.3	−0.8
	c	− 0.6	−1.9	− 0.4	+0.8	− 0.5	−0.6	+ 0.4	+ 0.4	− 0.5	− 0.1	−2.6	+ 1.0	−0.4	+0.4	−0.4	+ 0.3	+ 0.9	+ 2.2	+0.7	+1.0
April	a	+ 1.3	−0.5	− 0.5	+0.3	+ 0.2	+2.7	− 0.1	− 2.8	+ 0.2	0.0	−0.2	− 0.4	+1.0	+0.2	+0.2	− 0.2	− 0.4	+ 0.9	+0.2	+0.1
	b	− 1.9	−0.8	− 1.3	+0.5	− 0.9	−0.1	− 2.2	− 0.2	− 1.1	− 0.9	+0.7	− 1.2	−0.6	+0.9	0.0	− 1.2	+ 1.0	+ 2.1	+0.1	+0.5
	c	− 1.0	−1.0	0.0	−0.1	− 0.5	−0.5	− 1.3	+ 1.3	− 1.4	− 0.5	+1.0	− 0.8	−0.4	+0.3	0.0	+ 0.4	+ 0.7	+ 1.5	+0.2	+0.7
May	a	+ 0.8	−2.4	− 0.4	−0.7	− 0.7	−0.8	− 1.1	0.0	− 0.8	− 0.7	+1.8	− 0.7	+4.5	−0.2	+1.3	+ 1.9	+ 1.2	− 2.1	+0.5	+0.4
	b	− 0.1	−0.3	− 1.3	−0.1	− 0.4	−0.7	+ 1.8	0.0	− 1.1	0.0	+0.5	+ 0.8	+4.9	−0.1	+1.5	− 2.7	+ 1.4	+ 1.6	+1.0	+0.3
	c	− 1.9	+0.9	− 2.6	−0.3	− 1.0	−0.2	− 2.6	− 4.2	+ 0.2	− 1.7	−1.7	+ 1.0	−3.5	0.0	−1.0	− 0.6	− 2.2	+ 4.0	+0.9	+0.5
June	a	− 0.5	−1.2	− 1.5	−0.1	− 0.8	+0.1	+ 0.1	− 2.9	− 0.3	− 0.8	−1.9	+ 0.1	−0.4	+0.2	−0.5	+ 0.7	− 2.4	+ 4.2	+0.1	+0.7
	b	+ 0.3	−1.9	− 3.6	+0.6	− 1.2	−0.2	+ 0.5	− 3.9	− 0.1	− 0.9	−0.1	− 2.6	−3.4	+0.4	−1.4	− 0.4	− 1.6	+ 5.4	−0.1	+0.8
	c	− 0.2	+0.6	− 1.5	+0.1	− 0.2	−0.9	− 0.7	− 3.1	+ 1.3	− 0.8	0.0	− 2.0	+1.7	+0.2	0.0	+ 0.4	+ 0.1	+ 0.3	+0.5	+0.4
July	a	− 2.3	−0.3	0.0	+0.4	− 0.6	−0.1	+ 0.3	− 1.8	− 0.8	− 0.6	−0.4	0.0	+2.2	+0.5	+0.8	− 1.0	− 1.1	+ 1.6	+0.8	+0.1
	b	− 1.1	−0.3	− 1.8	−0.6	− 1.0	+0.1	0.0	+ 0.9	+ 0.4	+ 0.4	−0.5	− 1.4	+0.7	−0.6	−0.4	− 0.9	− 2.2	+ 2.1	+0.7	−0.1
	c	− 1.2	−0.6	+ 1.0	−0.2	− 0.2	+0.1	− 0.4	+ 1.1	0.0	+ 0.2	−0.6	− 0.6	−3.3	−0.9	−1.4	− 1.0	− 0.8	+ 2.4	−0.2	+0.1
Aug.	a	− 0.1	+0.1	− 3.0	−0.5	− 0.9	−0.6	0.0	− 2.9	+ 0.1	− 0.8	−0.9	− 1.3	−1.8	+0.6	−0.8	− 0.6	− 1.2	+ 6.8	−0.4	+1.2
	b	− 0.1	+0.7	− 1.9	+0.3	− 0.2	−0.3	− 0.1	− 0.6	+ 0.5	− 0.1	−0.5	− 0.9	+2.0	+0.8	+0.4	+ 0.7	+ 0.9	− 2.5	+0.1	−0.2
	c	+ 0.8	−0.7	− 1.9	+0.8	− 0.2	+0.1	− 0.2	− 4.3	− 0.9	− 1.3	−0.3	+ 1.7	−2.0	+0.4	0.0	+ 0.8	− 0.6	+ 0.3	−0.3	0.0
Sept.	a	− 1.3	+0.1	− 1.4	−0.9	− 0.9	−0.6	0.0	− 1.6	− 0.6	− 0.7	+1.3	− 2.2	−3.1	+0.8	−0.8	+ 0.6	− 0.1	− 3.7	−0.5	−0.9
	b	− 0.3	+0.8	− 0.5	+0.1	0.1	+1.2	− 1.7	− 3.6	− 0.1	− 1.0	+0.2	− 1.1	−1.4	+0.5	−0.4	+ 1.8	+ 1.6	− 1.2	−0.7	+0.4
	c	+ 0.7	+2.7	− 1.6	0.0	+ 0.4	−0.3	− 1.4	− 3.7	− 0.1	− 1.4	+0.1	− 0.3	+0.8	+0.2	+0.2	+ 1.3	− 0.6	+ 2.7	0.0	+0.9
Oct.	a	− 0.4	+2.7	+ 0.1	+0.8	+ 0.8	+0.9	− 0.7	− 2.1	− 0.9	− 0.7	+0.1	− 1.3	−0.8	+0.9	−0.3	− 2.0	+ 0.2	+ 0.6	+0.3	−0.2
	b	+ 1.9	−2.7	− 1.2	−0.6	− 0.7	−0.9	− 0.8	− 2.3	− 0.7	− 1.2	−0.3	+ 1.2	−2.5	+0.6	−0.2	− 1.0	+ 2.5	− 3.8	−0.6	−0.7
	c	− 1.4	−0.6	+ 1.7	0.0	− 0.1	−0.1	+ 0.9	− 2.5	0.0	− 0.4	+1.4	+ 0.9	−2.0	+0.4	+0.2	− 1.3	+ 0.9	+ 0.6	+0.4	+0.2
Nov.	a	− 0.1	+0.3	− 1.3	−0.8	− 0.5	+0.2	+ 0.1	− 2.8	+ 0.1	− 0.6	−1.0	+ 0.9	−0.6	−0.2	−0.2	+ 0.6	− 1.4	− 4.0	−1.1	−1.5
	b	− 0.5	+1.6	+ 0.9	−0.9	+ 0.3	−0.6	− 1.5	− 2.4	0.0	− 1.1	−1.3	+ 0.7	+1.1	−0.7	0.0	+ 0.8	+ 1.7	+ 0.2	−0.2	+0.2
	c	− 0.6	−1.6	− 0.2	−0.4	− 0.7	0.0	+ 0.1	− 1.2	− 0.2	− 0.3	+1.6	+ 1.6	+0.7	−0.4	+0.9	− 0.2	− 2.0	+ 0.1	0.0	−0.5
Dec.	a	+ 2.0	−1.7	− 2.8	−1.0	− 0.9	−1.7	+ 3.4	− 1.2	− 0.2	+ 0.1	+1.8	− 0.9	−0.8	−1.2	−0.3	− 2.5	− 0.5	+ 1.4	+0.2	−0.4
	b	+ 0.1	−5.7	− 1.4	+0.1	1.7	+0.9	+ 1.4	+ 3.4	− 0.3	+ 1.4	+2.0	− 1.0	−1.2	+0.5	+0.1	− 0.6	− 2.2	+ 1.2	+0.5	−0.3
	c	− 3.8	+1.4	+ 1.2	−0.3	− 0.3	+3.3	− 0.4	+ 1.2	− 0.3	+ 1.0	−2.3	+ 0.4	+1.4	+0.4	0.0	+ 1.8	− 4.2	− 0.8	+0.4	−0.7
Sum		−20.1	−9.5	−25.6	−0.9	−13.9	−8.1	−11.4	−47.9	−11.5	−19.4	+8.6	−24.0	−8.9	+2.2	−4.9	−23.1	−12.1	+22.7	−1.1	−3.5
$\Sigma\Delta^2$		85	96	97	8	18.49	71	61	187	14	27.02	191	90	149	7	18.44	136	98	226	8	25.00

TABLE XI. — *Continued.*

Simultaneous Departures of Temperature in Regions in ° C.

		1896					1897					1898					1899			
		U.S. I	U.S. II	Arg.	Sa-moa	Mean	U.S. I	U.S. II	Arg.	Sa-moa	Mn.	U.S. I	U.S. II	Arg.	Sa-moa	Mn.	U.S. I	U.S. II	Sa-moa	Mn.
Jan.	a	−2.0	0.0	− 1.0	+0.2	− 0.7	+0.6	−0.2	−1.7	−0.4	−0.4	−0.1	+1.8	+ 0.8	−1.1	+0.4	+0.1	−0.7	−0.4	−0.3
	b	−0.1	+ 2.3	− 0.2	−0.1	+ 0.5	+1.0	+0.6	+0.3	−0.5	+0.4	+3.0	−3.4	− 0.4	−0.5	−0.3	+1.2	+1.6	−1.5	+0.4
	c	+1.1	+ 2.8	+ 0.5	−0.6	+ 0.9	−4.9	+1.6	−0.1	+0.1	−0.8	+1.9	−3.0	+ 1.7	−0.3	+0.1	−0.6	+3.1	−0.5	+0.7
Feb.	a	+1.6	+ 0.8	+ 0.3	−0.4	+ 0.6	−1.5	+1.3	+0.5	−0.4	0.0	−2.0	+1.5	− 1.5	−0.2	+0.2	−4.0	−3.7	0.0	−2.6
	b	−1.6	+ 2.5	0.0	−0.8	+ 0.4	+1.2	−1.6	−0.5	−0.6	−0.4	+2.3	+1.0	− 0.2	−0.9	+0.6	−7.1	+0.9	−0.7	−2.3
	c	+0.5	+ 2.4	− 0.9	+0.1	+ 0.5	−0.1	−1.6	+1.2	+0.3	0.0	−0.9	+1.9	+ 3.0	−0.7	+0.8	+1.1	+0.5	....	+0.8
Mar.	a	−0.1	− 1.8	+ 2.6	−0.3	+ 0.1	+2.8	−2.4	+1.7	+0.8	+0.7	−0.1	+1.6	+ 1.7	−0.9	+0.6	−0.6	+1.2	....	+0.3
	b	−3.7	+ 1.3	− 0.8	+0.3	− 0.7	+2.7	−3.7	+3.3	+0.2	+0.6	+4.4	−3.3	− 1.6	−0.9	−0.4	+0.8	−1.6	−0.4	−0.4
	c	−0.6	+ 1.7	− 0.6	+0.2	+ 0.5	+1.6	−3.4	+3.1	0.0	+0.3	+0.5	−3.1	− 2.4	−0.7	−1.4	−0.4	−0.6	−0.1	−0.4
April	a	−2.5	+ 0.6	− 1.1	0.0	− 0.2	+0.5	−1.3	+1.2	+0.8	+0.3	−2.4	−0.2	+ 0.4	−0.3	−0.6	−3.1	+0.7	−1.4	−1.3
	b	+3.2	− 3.4	− 1.9	−0.3	− 0.6	−1.3	+2.2	+2.1	−0.1	+9.7	+0.2	+2.2	− 2.2	0.0	0.0	−0.6	+1.8	+0.4	+0.4
	c	+1.2	− 1.7	+ 4.1	−0.4	+ 0.8	+0.2	−0.1	+2.3	−0.4	+0.5	−0.7	+2.5	− 1.7	−0.7	−0.2	+0.6	−1.3	+0.7	0.0
May	a	+1.6	− 2.6	− 3.5	+1.1	+ 0.9	−0.6	+1.0	+3.6	−0.1	+1.0	−1.7	−0.8	− 0.3	−0.6	−0.8	+1.0	−3.1	−0.2	−0.8
	b	+2.3	− 3.3	+ 1.8	+0.4	+ 0.3	−0.3	+1.9	+3.2	0.0	+1.2	+0.6	−1.0	+ 1.8	+0.1	+0.4	+0.6	−1.2	0.0	−0.2
	c	+0.7	+ 1.7	+ 1.7	+0.3	+ 1.1	−1.7	+1.9	−2.1	−0.1	−0.5	+0.3	−0.8	+ 2.8	−0.5	+0.4	−0.1	−2.4	−0.3	−0.9
June	a	+0.3	+ 0.6	+ 1.5	+0.1	+ 0.6	−1.8	+0.2	+3.4	+0.5	+0.6	+0.3	+0.4	+ 2.1	+0.1	+0.7	+1.7	−2.7	−0.4	−0.5
	b	−0.8	+ 3.9	− 2.8	−0.3	0.0	+0.3	−0.4	+1.7	+0.6	+0.6	−0.1	+0.8	+ 4.6	−0.5	+1.2	−0.4	+0.8	−0.8	−0.1
	c	+0.1	+ 1.0	+ 0.8	−0.9	+ 0.2	−0.1	−0.5	−3.0	−0.3	−1.0	+0.2	+0.9	+ 0.5	0.0	+0.4	−0.9	+0.6	−0.7	−0.3
July	a	−0.8	+ 1.2	+ 5.2	−1.6	+ 1.0	+0.6	−1.0	−3.5	−0.3	−1.0	−0.1	0.0	− 3.6	+1.1	−0.6	−0.6	+1.2	+0.4	+0.3
	b	−0.6	+ 0.1	+ 2.0	−1.9	+ 0.1	−0.8	−0.3	−0.9	+0.3	−0.4	−0.1	+0.6	− 1.4	+1.0	0.0	−0.2	−0.2	+0.4	0.0
	c	+0.3	− 1.3	+ 3.3	+0.1	+ 0.6	+0.1	−1.5	−1.1	−0.2	−0.7	+0.7	+0.9	+ 0.4	+0.4	+0.6	−0.2	−0.3	0.0	−0.2
Aug.	a	+1.6	− 1.1	+ 1.9	....	+ 0.8	−0.2	−0.4	−0.6	−0.1	−0.3	−0.2	+0.2	− 1.2	−0.5	−0.4	+0.1	−2.4	−0.1	−0.8
	b	+0.2	+ 1.0	+ 4.8	....	+ 2.0	−0.7	−0.3	−2.7	+0.7	−0.8	−0.1	+3.0	− 4.3	−0.4	−0.2	+0.1	−1.9	−0.1	−0.6
	c	−0.3	+ 0.6	+ 5.1	0.0	+ 1.4	−0.1	+1.4	−2.0	−0.1	−0.2	+1.0	+0.4	− 0.7	+0.1	+0.2	+0.8	−0.1	....	+0.4
Sept.	a	−0.2	+ 0.7	+ 0.7	+0.6	+ 0.4	+0.3	+0.5	−2.4	−0.7	−0.6	+0.8	+0.9	− 1.2	+0.6	+0.3	+0.7	+0.6	+0.8	+0.7
	b	+1.0	+ 1.3	+ 1.7	+0.1	+ 1.0	+1.0	+0.2	+0.7	−0.4	+0.4	0.0	+2.1	− 3.0	+0.3	−0.2	−0.3	+0.9	0.0	+0.2
	c	−1.1	− 0.7	+ 4.8	+0.1	+ 0.8	−1.1	+2.7	−1.4	−0.3	0.0	+1.5	+1.2	+ 0.7	−0.4	+0.8	−1.6	+2.6	+0.5	+0.5
Oct.	a	−1.6	+ 1.5	− 2.0	+0.4	+ 0.4	+0.2	+0.5	+3.5	−0.9	+0.8	+2.8	−0.7	− 2.4	−0.2	−0.1	−1.7	0.0	+0.1	−0.5
	b	−1.6	+ 1.6	+ 2.0	+1.1	+ 0.8	+2.5	−1.0	+1.2	+0.3	+0.5	−0.7	+0.5	− 3.9	−0.3	−1.1	+2.8	−2.8	+0.1	0.0
	c	+1.5	− 0.3	+ 2.2	+0.2	+ 0.9	+0.7	−1.0	−1.8	0.0	−0.5	−2.8	+1.3	− 1.2	−0.2	−0.7	+1.7	+0.3	0.0	+0.7
Nov.	a	+0.8	− 1.6	− 1.0	−1.1	− 0.7	+0.9	−0.7	−2.5	+0.1	−0.6	+0.5	+1.8	− 1.3	−1.0	0.0	−1.5	+2.6	+0.8	+0.6
	b	+2.6	+ 3.2	+ 0.9	−0.4	+ 1.6	+0.9	+2.4	+2.0	−1.2	+1.0	−0.3	−0.6	− 3.5	−0.3	−1.2	+2.6	+1.8	+1.0	+1.8
	c	+2.1	− 1.7	+ 0.9	+0.3	+ 0.4	+0.8	+2.1	−1.2	−0.9	+0.2	−2.6	−1.7	− 1.7	−0.6	−1.6	+0.7	+1.7	−0.6	+0.6
Dec.	a	−0.3	+ 0.6	− 0.5	+0.2	0.0	+0.5	−1.8	+1.4	+0.3	+0.1	−2.9	−0.2	+ 0.8	+0.2	−0.5	+0.5	−0.1	−0.4	0.0
	b	+1.1	+ 1.4	− 2.3	+1.0	+ 0.3	+1.2	−2.5	+0.4	−0.5	−0.4	−2.2	−1.7	− 1.4	−0.1	−1.4	+1.7	−1.2	−0.2	+0.1
	c	−0.8	+ 2.4	+ 1.5	+0.3	+ 0.8	−1.0	−2.3	+0.6	+0.5	−0.3	+1.4	−0.8	+ 0.4	−0.3	+0.2	−2.6	+0.8	+0.2	−0.5
Sum		+5.1	+17.7	+33.9	−2.0	+16.8	+4.4	−7.5	+9.9	−3.0	+1.0	+2.4	+6.2	−16.4	−8.4	−3.8	−7.7	−2.6	−3.4	−4.2
ΣΔ²		77	129	206	12	21.94	70	93	148	5	13.70	95	99	163	7	17.00	124	101	8	22.96

Table XI.— Concluded.

Simultaneous Departures of Temperature in Regions in °C.

		1900				1901				1902				1903				1904			
		U.S. I	U.S. II	Samoa	Mn.	U.S. I	U.S. II	Samoa	Mn.	U.S. I	U.S. II	Samoa	Mn.	U.S. I	U.S. II	Samoa	Mn.	U.S. I	U.S. II	Samoa	Mn.
Jan.	a	−1.0	+3.0	+0.7	+0.9	+1.0	+0.7	+0.5	+0.7	−1.0	+3.6	+0.8	+1.1	−1.2	+1.9	+0.8	+0.5	−2.4	−0.8	......	−1.6
	b	+2.5	+4.3	−0.1	+2.2	+0.7	+2.7	+0.7	+1.4	−0.8	+2.3	−0.3	+0.4	−1.7	+0.3	+0.6	−0.3	−0.9	+1.6	......	+0.4
	c	−0.4	+2.2	+1.0	+0.9	+0.9	+2.9	+0.1	+1.3	−1.7	−3.0	0.0	−1.6	+2.3	+2.7	+0.8	+1.9	−2.2	−1.8	......	−2.0
Feb.	a	−0.1	+1.5	+0.8	+0.7	−1.4	−0.4	+0.6	−0.4	−4.6	−0.8	0.0	−1.8	+2.4	+4.4	+0.7	+2.5	+0.1	0.0	−0.1	0.0
	b	−2.6	+0.1	−0.1	−0.9	−1.0	+0.1	−0.4	−0.4	−4.0	+3.3	+1.2	+0.2	−1.1	−4.9	+0.4	−1.9	−3.3	+0.7	−0.5	−1.0
	c	−2.8	+2.6	+0.7	+0.2	−4.2	+5.2	+0.3	+0.4	−0.1	+0.9	+0.7	+0.5	−0.8	−1.3	+0.3	−0.6	−0.2	+3.3	−0.7	+0.8
Mar.	a	−0.2	+3.5	−0.9	+0.8	−0.6	+2.1	0.0	+0.5	−0.5	−0.2	−0.5	−0.4	+2.2	−0.8	+0.2	+0.5	+0.7	+2.3	0.0	+1.0
	b	−2.4	+2.9	0.0	+0.2	+0.2	+0.5	+0.6	+0.3	+0.6	−0.8	+1.2	+0.3	+3.6	−0.1	+0.8	+1.4	+0.6	+0.9	−0.2	+0.4
	c	−0.9	+1.6	+0.1	+0.2	+0.7	−2.0	−0.2	−0.5	+2.4	−3.4	0.0	−0.3	+0.3	+1.9	+0.9	+1.0	+1.1	−2.2	−0.4	−0.5
April	a	−1.0	−1.0	+0.2	−0.6	−1.4	−3.3	+0.8	−1.3	−1.7	+0.9	−0.3	−0.4	+1.3	−0.1	+0.7	+0.6	−0.1	+0.6	−0.4	0.0
	b	−0.8	+0.4	+0.1	−0.1	−2.4	−0.3	+0.4	−0.8	−0.8	+1.6	−0.1	+0.2	−0.8	−1.7	+0.2	−0.8	−2.1	+3.0	−0.2	+0.2
	c	+0.7	−2.4	−0.2	−0.6	−1.6	+2.3	+1.3	+0.7	+1.5	−0.7	+0.1	+0.3	−1.9	+0.7	−0.4	−0.5	−1.4	−2.3	−0.6	−1.4
May	a	−0.2	+1.4	+0.6	+0.6	−0.1	+0.5	+1.5	+0.7	+1.3	+1.2	+0.8	+1.1	+1.9	+0.9	+0.1	−0.3	−0.3	−0.2	+0.6	0.0
	b	+0.4	0.0	−0.4	0.0	−0.1	+2.6	0.0	+0.9	+0.7	−0.9	+0.5	+0.1	+0.1	+0.3	+1.1	+0.4	−1.7	+1.0	+0.1	−0.2
	c	−0.6	+1.9	−0.9	+0.1	−2.2	−0.4	−0.2	−0.9	+0.3	+0.4	+0.4	+0.4	−0.1	−2.0	−0.7	−0.9	+0.4	−0.1	−0.1	+0.1
June	a	+0.1	+0.1	+0.2	+0.1	−0.9	−0.6	−0.4	−0.6	+0.4	+0.7	+0.2	+0.4	−1.4	+1.9	+0.7	+0.4	−0.2	0.0	+0.6	+0.1
	b	−1.0	+0.8	−0.8	+0.3	−0.5	−1.7	−0.7	−1.0	+0.6	+0.3	+0.1	+0.3	+3.0	+0.6	+0.6	−0.6	−1.4	+1.8	+0.2	+0.2
	c	+0.4	+3.2	−1.1	+0.8	+1.3	+0.5	+0.5	+0.8	−1.7	+1.5	−0.3	−0.2	+1.7	+0.7	+0.4	−0.2	−0.9	+2.0	−1.0	−0.6
July	a	+0.3	−0.4	−0.1	+0.1	+0.6	+1.0	+1.5	+1.0	+0.2	−3.5	+0.3	−1.0	+0.2	−0.8	+0.2	−0.1	−1.0	+0.3	−0.2	−0.5
	b	+0.1	+0.8	−0.5	+0.1	+0.2	+1.9	−0.2	+0.6	−0.2	+0.2	+0.2	+0.1	+0.9	+0.6	+0.1	−0.1	−0.8	−0.4	−1.1	−0.8
	c	−0.8	+0.7	−0.5	+0.2	+0.2	+0.9	+0.3	+0.5	−0.7	+0.2	+0.4	0.0	−0.6	−0.6	+0.2	−0.3	−1.9	−0.8	+0.2	−0.8
Aug.	a	+0.1	+0.4	+0.7	+0.4	−0.1	+0.7	+0.7	+0.4	+1.1	+1.0	−0.1	+0.3	−1.4	−0.6	+0.6	−0.5	−1.3	+0.4	−0.7	−0.5
	b	+1.5	−0.8	......	+0.3	+0.4	−0.4	+0.3	+0.1	−0.8	−1.0	+0.4	−0.5	−1.3	+1.4	+0.0	0.0	−1.2	+0.4	−0.3	−0.4
	c	+1.2	−0.2	+0.1	+0.4	−0.1	+1.4	−1.2	0.0	−0.2	+0.6	−0.9	−0.2	+0.7	+1.6	+1.0	+1.1	−0.9	+0.3	−0.4	−0.3
Sept.	a	+1.8	+0.3	+0.6	+0.9	−0.5	−0.3	+0.2	−0.2	−0.3	+1.6	+0.1	+0.5	−0.2	+1.2	+0.5	+0.5	−0.3	+1.8	−0.3	+0.4
	b	−1.2	−0.7	+0.1	+0.2	−0.6	−0.3	+0.1	−0.1	−2.4	+0.1	−0.8	−1.0	+0.4	−2.2	−0.8	−0.9	+0.3	+1.7	−0.4	+0.5
	c	+2.4	−2.2	+0.9	+0.4	−0.3	−0.7	0.0	−0.1	+1.7	−0.6	−0.3	+0.3	−0.8	+0.6	−0.6	−0.3	+0.9	−0.1	0.0	+0.3
Oct.	a	+1.9	+0.4	0.0	+0.8	−0.7	+0.2	−0.2	−0.2	−0.1	−0.1	+0.2	0.0	+0.9	−0.6	+0.7	+0.3	+1.2	+1.7	−0.2	+0.9
	b	+0.4	+1.6	+0.5	+0.8	−0.5	+1.8	−0.5	+0.3	+0.9	+1.6	−0.3	+0.7	−0.8	+1.3	0.0	+0.2	+0.2	−0.2	+0.2	+0.1
	c	+3.5	−0.5	+0.5	+1.2	+1.3	+2.0	−0.3	+1.0	+1.2	+2.4	+0.3	+1.3	−1.5	+1.4	−0.1	−0.1	−1.7	+1.8	+0.2	+0.1
Nov.	a	+0.9	+3.7	−0.2	+1.5	−0.8	+2.2	+0.5	+0.6	+2.0	+2.0	+0.6	+1.5	+0.2	+0.9	−0.5	+0.2	−0.8	+2.4	−0.4	+0.4
	b	+0.5	+2.8	−0.1	+1.1	−2.1	+1.4	−0.5	−0.4	+4.0	0.0	+0.2	+1.4	−0.3	+0.8	−0.2	+0.1	−1.5	+1.6	+0.2	+0.1
	c	+2.6	+1.8	+0.1	+1.5	−0.9	+3.8	−0.1	+0.9	+1.8	−1.6	+0.6	+0.3	−3.7	+3.1	+0.2	−0.1	+0.3	+3.7	0.0	+1.3
Dec.	a	−0.7	+3.3	+1.0	+1.2	+0.6	+2.1	+0.1	+0.5	−0.6	+0.6	+0.5	+0.2	−3.3	−0.6	−0.2	−1.4	−1.3	+0.4	0.0	−0.3
	b	−0.3	+1.6	−0.3	+0.3	−3.9	+2.6	+0.6	−1.8	+1.0	−1.8	+0.2	−0.2	−2.4	−0.1	−0.5	−1.0	−3.6	+1.4	−0.1	−0.9
	c	+1.3	−0.3	−1.4	0.1	−0.7	+1.0	+0.4	+0.2	−1.9	+0.9	+1.0	0.0	−0.8	+0.2	+0.2	−0.1	+0.4	−0.1	+0.5	+0.3
Sum		+8.0	+38.4	+1.3	+15.9	−21.1	+28.1	+7.1	+4.9	−3.4	+9.5	+7.1	+4.3	−19.2	+13.0	+9.0	+0.6	−27.2	+23.7	−5.5	−4.2
$\Sigma\Delta^2$		73	13.6	10	21.93	68	131	8	19.90	100	100	7	19.05	99	100	7	26.14	68	88	2	18.27

What we have next to do is to sum all the squares through the whole period of 33 years. This summation, with the partial values of Δ which result from it, is shown in the next table. The most noteworthy circumstance here brought out is the complete absence of any systematic value of the residual Δ. This may be shown by dividing the series into three parts during each of which the stations remained unchanged. The result is as follows:

Summation of Squares for Ten-day Deviations.

$n = 3$

Years	U. S. I Σv_1^2	U. S. II Σv_2^2	Arg. Σv_3^2	$\Sigma \tau^2$	$n^2 \Sigma \tau^2$	Σv^2	Δ
1872	120	44	97	41.4	373	261	+112
73	62	48	48	12.9	116	148	— 32
74	70	64	102	18.2	164	236	— 72
75	114	96	81	24.0	216	291	— 75
76	124	58	96	37.6	338	278	+ 60
77	76	81	113	41.2	371	270	+101
78	84	89	97	40.8	367	270	+ 75
79	101	98	63	32.9	296	262	+ 36
1880	172	149	109	51.5	464	430	+ 36
81	104	113	70	31.7	285	287	— 3
82	91	114	62	19.8	178	267	— 90
83	40	99	108	28.7	258	247	+ 12
84	76	87	115	19.5	176	278	—102
85	64	85	120	30.0	270	269	0
86	80	154	85	25.4	229	319	— 90
87	83	58	119	31.9	287	260	+ 27
88	64	131	136	32.2	290	331	— 39
89	80	116	93	29.3	264	289	— 24
Sum	1605	1674	1714	549.0	4942	4903	— 66

$n = 4$

Years	U. S. I Σv_1^2	U. S. II Σv_2^2	Arg. Σv_3^2	Samoa Σv_4^2	$\Sigma \tau^2$	$m^2 \Sigma \tau^2$	Σv^2	Δ
1890	126	127	132	7	27.4	438	392	+ 48
91	69	96	100	5	11.1	178	270	— 96
92	85	96	97	8	18.5	296	286	+ 8
93	71	61	187	14	27.0	432	333	+100
94	91	90	149	7	18.4	294	337	— 40
95	136	98	226	8	25.0	400	468	— 68
96	77	129	206	12	21.9	350	424	— 72
97	70	93	148	5	13.7	219	316	— 96
1898	95	99	163	7	17.0	272	364	— 92
Sum	820	889	1408	73	180.0	2879	3190	—308

$n = 3$

Years	U. S. I Σv_1^2	U. S. II Σv_2^2	Samoa Σv_3^2	$\Sigma \tau^2$	$n^2 \Sigma \tau^2$	$n^2 \Sigma \tau^2$	Σv^2	Δ
1899	124	101	8	22.2	67	200	233	—33
1900	73	136	10	21.9	66	197	219	—21
01	68	131	8	19.9	60	179	207	—27
02	100	100	7	19.0	57	171	207	—36
03	99	100	7	26.1	78	235	206	+27
04	68	88	2	18.3	55	165	158	+ 6
Sum	532	656	42	127.4	383	1147	1230	—84

It is not necessary to compute the value of τ_0 from these data because it is evidently evanescent, the mean coming out with an imaginary value. In fact the values of Δ as they come out in the last columns of the table are less than their probable errors by amounts smaller than could be expected, except as the result of chance. There is therefore no evidence of any irregular fluctuation having a period between ten days and several years.

§ 15. *Search for Variations Synchronous with the Sun's Synodic Rotation by the Method of Time-correlation.*

Granting the existence of variations in the solar constant it is extremely improbable, and indeed almost inconsistent with any theory of what is going on in the sun, to suppose them to take place simultaneously over the entire photosphere. We should expect them to be mostly confined in each case to some limited region ; then, when this region became visible from the earth, we should experience a change in the solar heat, which would reach its maximum or minimum when, in consequence of the sun's rotation, the meridian of the hot or cool region of the photosphere passed the middle of the sun's disc as seen from the earth. After this the effect would diminish, and would disappear entirely as the region disappeared from our sight on the sun's western limb, to be renewed when it reappeared on the eastern limb. Thus we should have a fluctuation in the terrestrial temperature having the period of the sun's synodic rotation.

Were the period of the rotation a well-defined constant, and were the excess of temperature in any region of one hemisphere permanent, the effect could be determined in the same way that we have determined that of the solar spots, by forming equations of condition for the coefficients expressing the amplitude of the resulting fluctuations. But there are two conditions which would render this method illusory. The first is that, owing to the different periods of rotation in different parallels of solar latitude, there would be no one invariable period of the phenomenon. The other impeding condition is that we must expect such deviations of temperature within any region of the sun to be temporary, lasting only a few weeks or months, and then disappearing, to reappear in some other region of the sun. Then the effect would appear entirely non-periodic if followed during long intervals of time, and could be detected only by the statistical methods already developed. If the change of solar temperature ordinarily disappeared before a rotation was completed, the effect would be entirely irregular and non-periodic. But if it continued through one or more solar rotations, as would probably be the case, then the effect would be temporary fluctuations of temperature having the period of the synodic rotation, but changing their epoch from time to time, and thus annulling each other if we treated them as continuous through long periods of time. We have shown how a phenomenon of this kind can be detected, even if it lasts in each special case through little more than a single rotation of the sun, by the method of time-correlation. The following considerations may guide our course of thought on the subject.

Let us grant that on any occasion a region of the sun extending to, at least near the equator, and hotter than the photosphere in general, is carried past the apparent solar meridian by the sun's rotation. During a period of ten days it will be sufficiently near the meridian to produce a rise in terrestrial temperatures. Then, as it disappears, the temperature will begin to fall until the region reappears on the sun's eastern limb. Then there will be another rise in the temperature, showing a rhythmical movement of the latter. What we have to do is to inquire into the fluctuations of temperature with a view of determining whether there can be found any rythmical tendency among them to recur at the end of about 26 days. This is most completely and rigorously done by searching for correlations between terrestrial temperatures at any one epoch, or through one term, and during the following terms up to 26 days or more. To discover the effect it seems desirable to take terms as short as five days, and to carry their study continuously forward. It is then certain that, if any exceptionally hot or cool region of the photosphere has been carried past our solar meridian, the effect will be at its maximum during at least some one term. A study of the temperatures during the five terms following will then show what changes in terrestrial temperature have taken place while the special region was moving around and returning again to the solar meridian.

I have chosen for this research the temperatures at San Diego because they are fairly steady, and it chanced that the data for 5-day terms were available through a period of more than 30 years, and therefore nearly 400 synodic rotations of the sun. The research was confined to this station more through practical considerations than because it was absolutely the best. If the clearest result is to be brought out, stations in some continental interior, where the temperature is little affected by the ocean, and where the irregular fluctuations are as small as possible, should be preferred. Moreover, as the effect sought for is common to the whole globe, the mean of the largest practical number of such stations should be used. But the writer conceives that a fairly certain result can be derived from San Diego alone.

The method by which the periodicity is to be detected is that developed in § 2. We take the departures of temperature during a number of consecutive five-day terms, as great as we choose. In the present case we have chosen six, making a period of thirty days. The departures during the six terms of this period are designated as

$$a_0, \; a_1, \; a_2, \; a_3, \; a_4, \; a_5$$

Beginning with the first five-day term, we now multiply a_0 into each of the following five departures, and write their products in a horizontal line. A new period is then begun with a_1 of the preceding term so that the departure which appears as a_1 in the first line becomes a_0 in the second, after which it is not used. Thus each individual departure enters into six consecutive periods.

It does not seem necessary to encumber the work by giving the individual departures, 2376 in number, in detail. The following commencement of the table will show how the individual products were formed.

It being usual to designate the ten-day terms of each month as a, b, and c, we designate the five-day terms as a_1, a_2, b_1, etc.

The column a_0 of the table gives the five-day departures of the normal temperature as determined from the records of the Weather Bureau. The method by which the six products in each line are formed will be readily seen, as the factors are all given for the first two lines, and can be readily understood for the lines following.

1872		a_0	a_0^2	a_0a_1	a_0a_2	a_0a_3	a_0a_4	a_0a_5
Jan.	a_1	−3.0	9.0	+4.2	+7.8	− 4.5	−0.9	−12.0
	a_2	−1.4	2.0	+3.6	−2.1	− 0.4	−5.6	+ 0.4
	b_1	−2.6	6.8	−3.9	−0.8	−10.4	+0.8	− 4.2
	b_2	+1.5	2.2	+0.4	+6.0	− 0.4	+2.4	+ 1.2
	c_1	+0.3	0.1	+1.2	−0.1	+ 0.5	+0.2	+ 0.2
	c_2	+4.0	16.0	−1.2	+6.4	+ 3.2	+2.0	− 5.2
Feb.	a_1	−0.3	0.1	−0.5	−0.2	− 0.2	+0.4	+ 0.8

Instead of showing at once the sum total, the addition has been grouped, the period of 33 years being divided into terms of 5 years each, except the last, which includes only 3 years. The results of the separate summations are as follows:

Period	$[a_0]^2$	$[a_0a_1]$	$[a_0a_2]$	$[a_0a_3]$	$[a_0a_4]$	$[a_0a_5]$
1872–76	2029	906	592	333	235	211
1877–81	2810	1506	660	895	897	921
1882–86	2664	1209	652	584	616	500
1887–91	2891	1249	716	416	625	732
1892–96	2790	1032	568	346	313	434
1897–01	2655	1141	673	487	505	547
1902–04	1639	826	570	492	353	275
Sum	17478	7863	4431	3553	3544	3620
x_i		+0.450	+0.254	+0.205	+0.203	+0.207

In the bottom line of the table are given the coefficients of correlation found x_i by dividing the several sums of the products in the last five columns by the sums a_0^2.

The values of x thus found may be regarded as non-periodic. Were there any tendency toward a recurrence at the end of 25 days there should be a marked increase in the values of the 4th and 5th products, because the 5th corresponds to a completion of the sun's synodic rotation. It is true that there is a minute increase of 0.004 between the 4th and 5th terms of the set. But an examination of the several separate sums through which this is formed shows that the increase is too small and uncertain to be regarded as the effect of periodicity.

But a quasi-periodicity is still possible, the persistently positive sign of x indicating a tendency of the departures to persist through a period of more than 25 days. The exact general fact brought out by the correlation is as follows:

Whatever be the departure of temperature at San Diego during any 5-day term we may expect the subsequent departures to lie in the general average in the same direction for more than a month, the ultimate amount at the end of the month being about one fifth that of the departure taken as the initial one. This persistence certainly seems singular, and it may be that had the correlation period been extended, periodicity would have been brought out.

As a further illustration of the method, without expecting to reach definitive results, I have made a similar time-correlation of the general mean temperatures for each decade as given in Table XI. preceding. The correlation-products were carried through periods of four terms, or 40 days each, counting from the middle of the initial to the middle of the last term. The actual period included is 50 days between extremes. The result, summed by terms of three years, is as follows:

Years	$a_0{}^2$	$a_0 a_1$	$a_0 a_2$	$a_0 a_3$	$a_0 a_4$
1872–1874	65.4	+ 13.2	+14.1	+ 8.2	+ 4.2
1875–1877	103.7	+ 36.7	+10.5	+25.8	+11.5
1878–1880	124.7	+ 51.9	+19.0	+26.3	+33.3
1881–1883	80.6	+ 12.2	+ 0.4	− 5.6	− 5.3
1884–1886	75.4	+ 8.5	+ 0.4	+10.9	− 9.4
1887–1889	95.6	+ 29.2	+ 6.2	− 8.8	−10.3
1890–1892	55.5	+ 19.4	+11.3	+10.4	+ 7.3
1893–1895	68.6	+ 25.7	+16.4	+13.6	+12.2
1896–1898	53.4	+ 18.5	+11.4	+16.8	+ 7.7
1899–1901	66.7	+ 25.8	+ 0.1	+ 1.7	+10.1
1902–1904	64.4	+ 12.3	− 0.5	− 2.9	+ 2.2
Sum	854.0	+253.4	+89.3	+96.4	+63.5
x_i		0.296	0.105	0.113	0.074

A general tendency is here shown in the departures of temperature to continue in the same direction for a period of at least 50 days. The time required for them to disappear entirely can be determined only by continuing the products through a longer period, which requires little more than a work of routine computation.

What is striking in the present case is the small increase of the fourth sum, following the rapid diminution of the first three sums. This is what we should expect from temporary inequalities in the temperature of the two solar photospheres. If this is really the case we may estimate the change in question as affecting terrestrial temperatures by two- or three-hundredths of a degree. A more exhaustive inquiry nto this question certainly seems of scientific interest, but I must, as with the continuation of the present work generally, leave this in other hands. The main point reached is that the influence of any such inequality in the sun upon meteorological phenomena is so nearly evanescent that it can be brought out only by the most refined methods of investigation, and cannot be of practical import.

CHAPTER VII.

DISCUSSION OF RESULTS.

§ 16. *Summary of Conclusions.*

The general results of the preceding discussion, so far as concerns fluctuations in the sun's radiant energy, may be summed up in the following propositions.

1. A study of the annual departures of temperature over many regions of the globe in equatorial and middle latitudes shows consistently a fluctuation corresponding in period with that of the solar spots. The maximum fluctuation in the general average is $0°.13$ C. on each side of the mean for the tropical regions. The entire amplitude of the change is therefore $0°.26$ C., or somewhat less than half a degree of the Fahrenheit scale. As this fluctuation has ample time to produce its entire effect on the earth, we conclude from it that the corresponding fluctuation in the sun's radiation is 0.2 of one per cent. on each side of the mean.

2. Additional to this periodic fluctuation there is some rather inconclusive evidence of changes requiring generally about six years to go through their period, which can be most plausibly attributed to corresponding changes in the sun's radiation. The phenomena may be expressed in the briefest way by saying that, during the years 1871–1904, there seem to have been periods of two, three or four years warmer than the normal, followed by similar periods which were cooler than the normal. But although the general tendency is toward changes in this period of about six years, they show no such correspondence with the solar spots as justified their being attributed to the sun-spot period. Moreover, they do not appear in any marked way before 1871. The average departure from the mean being less than $0°.10$ C. prevents a more exact statement of their law, and still leaves open the question whether they are real. This can be settled only by a more complete discussion of meteorological data than the writer has attempted to make.

3. Apart from this regular fluctuation with the solar spots, and this possible more or less irregular fluctuation in a period of a few years, *the sun's radiation is subject to*

no change sufficient to produce any measurable effect upon terrestrial temperatures. The only admissible changes are such as going through their period in 10 days or less, would produce no effect upon 10-day mean departures. Whether any such fluctuations exist, except those arising from the irregular changes of the spots and faculæ, is a question to be judged by the probabilities of the case.

4. There is a certain suspicion, but no conclusive evidence, of a tendency in the terrestrial temperature to fluctuate in a period corresponding to that of the sun's synodic rotation. If the fluctuations are real they affect our temperatures only a small fraction of one tenth of a degree.

5. To facilitate the criticism of the preceding conclusions, and their comparison with those reached by other investigators, we must point out what may be considered a limitation upon their scope. A careful study of the statistical method developed in § 4 will show that the primary intention is not to determine specific fluctuations, and attribute them to changes in the sun's thermal radiation, but only to find a general criterion for determining whether, as a general rule, the fluctuations have any other cause than the accumulation of accidental vicissitudes of temperature in the regions studied. Repeating once more in a condensed form the fundamental principle itself; when we determine the mean temperature of the globe by comparing the actual with the normal temperature at a great number of places through a number of time-terms, — we may determine the general world fluctuation by taking the mean of the departures in the separate regions during this term. This world-departure will have a certain probable deviation, arising from the probable deviations of the individual departures, the magnitude of which is easily computed.

If the world-departures are in general markedly greater than this probable deviation, we should have no difficulty in concluding that at the times of the greater departures the solar radiation was probably greater or less than the normal. Now the statistical method here applied is not intended to solve this easy problem should it arise (which it does not), but the more difficult one which arises when the actual departures do not ordinarily exceed their probable value, and when therefore we must be in doubt as to their arising from a cosmical cause. No sound method of research will enable us to formulate a conclusion on insufficient data, and the logically best method is that which will enable us to formulate all the conclusions that can be drawn. In the present case this is shown to be the probable value, during each time-term, of the square of a certain quantity τ_0 expressive of the increment of the solar radiation during that term. This quantity will have its probable accidental error, and therefore, if its objectively true value is evanescent, may still come out with a certain value, which is then as likely to be negative as positive. Having found this value through all the various terms, if the total sum comes out with a positive value markedly exceeding its probable value, we may infer with a corresponding degree of probability that some at least of the departures are real. If the excess is not great, then what we should conclude is that there is a greater or less probability, in a general way, that the sun's radiation is variable, but not that it had a definite variation at a definite time. To draw the latter conclusion from the data would be fallacious, not from any defect in the method but from the very nature of the case. But, if the well marked excess were a general rule, then we could fairly infer that, as a general rule, the fluctuations of temperature indicated corresponding fluctuations in the solar energy. For example, referring to the column τ in Table VI, which shows the residual departure of the annual temperatures after eliminating the effect of the sun-spot period, we may say that the temperature appears to have been above the normal in 1871, again in the years 1881–83 and again in 1896–97. It seems to have been below the normal in 1874–84, 1887 and 1892, and 1893.

Although these fluctuations, even if real, are so small that we cannot expect to trace them in any other meteorological phenomena than the temperature, the question of their reality is of scientific interest. This can be determined only by more extended researches.

To state the limitation in a more condensed form, the proof of general invariability does not positively establish the negative proposition that the sun's heat has never, on any one occasion whatever, undergone a perturbation during the period covered by our researches. In the absence of better positive evidence than is yet available, the assumption of such a perturbation would be a purely gratuitous one, to be refuted by a consideration of its improbability rather than by positive evidence.

§ 17. *Relation Between the Solar Radiation and Meteorological Processes.*

The preceding studies being primarily of fluctuations in the temperature of the air at the earth's surface, the question arises how far, from the steadiness of temperature we have established, we are justified in affirming that the sun's thermal radiation is steady in a corresponding degree. The consideration of this question will be facilitated by calling to mind certain points bearing upon it. A general proposition which, the writer conceives, needs no enforcement, is that so far as our science can show, the earth receives an appreciable supply of heat only from the sun. We may safely assume that the minute amount of heat reaching the earth's surface from the stars or other bodies in the celestial spaces, or by conduction from the earth's interior, is too minute to materially affect the temperature around us. This temperature is determined in a general way by the condition that it is such that the earth shall radiate into space as much heat as it absorbs from the sun's rays.

The radiations which reach the earth or its atmosphere from the sun are of two great classes. We have first *radiance* properly so called by which I understand radiant energy in its ordinary acceptation. This includes not only the rays commonly called *light*, but all other rays of the same class which differ from light only in wave length. It may here be remarked, parenthetically, that the use of the word "light" in physics is rather unfortunate, since the distinction of light and dark rays is not an objective one, but rests only upon the property of affecting the optic nerve. Thus, when we use the word light, we have one word for radiance between certain limits of wave length and no special term for radiance of the same kind of wave length without the visible limits.

Besides radiance as thus defined, we have abundant evidence that the sun sends, at least to the confines of our atmosphere, certain emanations which affect the magnetic needle, and which do not reach us in a steady stream, but fitfully, at irregular intervals. These emanations have, up to the present time, eluded all direct investigation. They are made known only by their effect upon the terrestrial magnetic force, as shown by magnetic storms. It therefore seems probable that those which reach the atmosphere are entirely absorbed in its outer envelopes.

The preceding study is practically limited to radiations of the first class. It is still questionable whether the magnetic or radio-active emanations, whatever they may be, appreciably affect the temperature. The recent researches of Maunder seem to show that they come mainly from the solar spots. Now, it is known that the radiance from the spots is less than from the rest of the photosphere. It follows that, if the emanations in question convey an appreciable amount of thermal energy, it does not reach the earth, but is absorbed in the upper regions of the air, perhaps almost at the surface of the atmosphere itself. But, were this the case, the extreme rarity of the air at high altitudes would result in a proportionately greater rise of temperature through a given radiation of thermal energy. In a word it seems highly improbable that emanations having radiant energy in considerable quantities could be absorbed by so rare a medium as the air at great heights above the earth.

The evidence afforded by the frequency of magnetic storms shows that the emanations in question are greatest at the period of sun-spot minimum when the terrestrial temperature is least. This affords an additional ground for believing that the thermal effect of the magnetic radiation is too small to produce any directly observable meteorological effect.

So far as research has yet gone, the balance of evidence would seem to favor the view that the phenomena of atmospheric electricity, especially of thunder storms, so far as they are changeable, arise mainly from terrestrial causes, and are but slightly influenced by solar emanations. Still, the question whether there is any relation between magnetic storms, which afford us the best available evidence of the emanations in question, and thunder storms or other exhibitions of movements of atmospheric electricity, is an interesting one, well worthy of investigation by rigorous statistical methods, and offering no difficulty. The main point to be enforced in the present connection is that our investigation includes the effect of all cosmical causes affecting the terrestrial temperature, and therefore of the emanations in question so far as they produce any thermal effect.

Dropping the consideration of magnetic, electric or radio-active emanations as belonging to another branch of the subject, because they do not cause appreciable fluctuation in terrestrial temperatures, we return to the main question now under consideration — that of the relation between fluctuations in the sun's thermal radiation and the corresponding changes in temperature. Accepting the fourth-power law of radiation, fluctuations in the general temperature of the globe of $0°.2$ C. on each side of the mean would produce corresponding changes of 0.3 of one per cent. in the radiation of heat by the earth into space. We have found that the fluctuations of world-temperature, if any at all occur, which is doubtful, do not exceed $\pm 0°.20$ C. We may therefore assign three tenths of one per cent. as the ordinary limit of fluctuation of the sun's radiation in lower periods. But the lag of temperature behind insolation is to be considered in the case of short periods.

Speaking in a general way, it is an observed fact that the maxima and minima of temperature in the temperate regions do not occur until about a month after the maxima and minima of radiation. But, admitting that a month will be required to produce the completed effect through the entire atmosphere and on the surface of the ground and ocean, it does not follow that the effect would be negligible in a shorter period. It is also an observed fact in regions of middle latitude that the rays of the sun between its rising and 2 p. m. elevate the temperature of the air at the earth's surface as read by the thermometer, by an amount ranging from 8° to 10° C. every day. Now, to fix the ideas, suppose that the sun's thermal radiance were increased by one per cent. of its whole amount through ten consecutive days. The result would be that the daily rise would be increased by an amount between $0°.06$ and $0°.10$. This rise would be in part lost during the night by increased radiation and transmission to the earth and upper air. But, as the earth and air grew warmer day after day the loss would be smaller and smaller, while the gain would continually accumulate. It follows that we should not have to wait more than a week for the change of one per cent. in the sun's energy to produce an effect exceeding that which our study of temperatures shows can be actually found in the world-temperature. But this does not preclude the possibility of much larger fluctuations in shorter periods, because it would take time for temporary increase in the sun's radiation to produce its full effect. The shorter the time that we suppose an increase or decrease to last, the greater it must be. It is mainly a question of judgment and probabilities whether changes of such very short period in the radiation can exist. The probabilities against them are based mainly on the fact that it is scarcely conceivable that any cause affecting the totality of the sun's radiation should act simultaneously over the entire photosphere. The most plausible cause of such fluctuations would be looked for in the faculæ and spots. These, and the phenomena connected with them are mainly local, never covering any important fraction of the sun's disc.

A yet more plausible source of change is found in possible fluctuations in the transparency of the solar envelopes. But these would take a long time to extend themselves over the entire photosphere. By allowing them a period of several weeks to spread over the sun, we bring them within the range of the present studies which then seem to establish their non-existence, except within the limits already several times mentioned.

A collateral question which is not included in the present research is whether the conclusions which have been drawn as to the constancy of the sun's radiation can be applied to other meteorological changes than those of temperature. The writer conceives that fluctuations of temperature are the primary cause of changes in precipitation, rainfall or great movements of the air, and fluctuations of the barometer. Confining ourselves within the limit of reasonable probability, the totality of rainfall must in the long run balance the evaporation. The rate of evaporation is, so far as is known, not influenced by electrical or magnetic conditions of any kind, but dependent solely upon the temperature and physical condition of the evaporating surface, and the temperature and motion of the air in contact with it. If the motions of the air are not affected by changes in the sun's radiation it would therefore follow that the rate of evaporation is determined solely by terrestrial conditions. This being granted it follows also that the total rainfall is determined in the same way. The total mass of the atmosphere being a constant, the integrated barometric pressure through the whole globe must also be a constant. Fluctuations in its amount in any region must therefore be balanced by opposite fluctuations in other regions and must be due to motions of the air which are determined only by conditions of temperature. If these views are correct it follows as the final result of the present investigation that *all the ordinary phenomena of temperature, rainfall and winds are due to purely terrestrial causes and that no changes occur in the sun's radiation which have any influence upon them.*

§ 18. *Comparison with Results of Langley's Work of 1903.*

Although the writer deems it appropriate for the most part to leave the farther discussion of his results, and their comparison with the views of others, to other investigators, an exception may well be made in the case of the very suggestive paper of Langley.* It should be premised that Langley does not present his results as conclusive, but only as showing seeming correlations between temperatures and bolometric measurements of the sun's radiation, the results of which should be tested by further researches. He gives the following general summary of his conclusions :

" A series of determinations of the solar radiation outside the atmosphere (the solar constant), extending from October, 1902, to March, 1904, has been made at the Smithsonian Astrophysical Observatory under the writer's direction.

" Care has been exercised to determine all known sources of error which could seriously affect the values relatively to each other, and principally the varying absorption of the Earth's atmosphere. Though uncertainty must ever remain as to the absorption of this atmosphere, different kinds of evidence agree in supporting the accuracy of the estimates made of it and of the conclusions deducted from them.

" The effects due to this absorption having been allowed for, the inference from these observations appears to be that the solar radiation itself fell off by about 10 per cent., beginning at the close of March, 1903. I do not assert this without qualification, but if such a change in solar radiation did actually occur, a decrease of temperature on the Earth, which might be indefinitely less than 7°.5 C., ought to have followed it."

The present writer understands that not only Langley's work with the bolometer, but observations by actinometric methods showed a remarkable diminution of the solar radiation, extending from some time in 1902 through a considerable portion of 1903. But as such observations are made only on the radiation which reaches the earth's surface the results still leave open the question whether the change was in the sun itself or was caused by increased absorption in the earth's atmosphere. The apparent diminution during the period in question has been plausibly attributed to the absorbing matter thrown up by the eruption of Mount Pelée on May 8, 1902. If the diminution of radiation was only apparent, being due to the absorption, we cannot, in the present state of science, decide whether there would be any effect upon terrestrial temperatures. While less heat would reach the earth directly, more would be absorbed in the middle regions of the atmosphere ; and this would apply both to the absorption of the sun's rays and of the heat radiated from the solid earth. The vapors of Mount Pelée, if they had any influence whatever, might, so far as we know,

* "On a Possible Variation of the Solar Radiation and Its Probable Effect on Terrestrial Temperatures," *Astrophysical Journal*, June, 1904.

have resulted in either a rise or a fall of the temperature of the earth in general. Hence even if we accept as unquestionable the correctness of the bolometric measures, it does not follow that there would be any corresponding change in the terrestrial temperature.

But Langley has brought forth what seems to be very strong evidence of a correlation between the temperatures in widely separated regions of the globe, using a method identical in principle with that of the present work, but including only the year 1903. His material was derived from the *Dekadenberichte* of the *Deutsche Seewarte* which gives ten-day temperatures in a great number of regions in various parts of the globe. The latter was divided by Langley into seven great regions and the mean departure found in each, on the same general plan that has been followed in the present work. The fluctuations in the seven regions were expressed in the usual way by curves, from a study of which the conclusion that there was a marked synchronism between the curves of temperature *inter se* seems quite plausible. The bolometric measures suffered so many interruptions that the curve representing them is frequently doubtful but, so far as it can be compared, there seems to be some correspondence between it and the temperature curves. Yet, the method of eye estimates through curves is one in which there is too much room for bias, and which does not admit of sufficient precision of determination. The correlation thus exhibited is quite at variance with the general conclusions of the present work, though these would not preclude the possibility of a marked chance correlation through any one year. But even for the special year 1903, it will be seen that the criterion of correlation is only

$$\Delta = 78 - 69 = 9$$

which does not rise above the expected result of chance accumulation of accidental deviations.

In view of the fact that, in the present work, the year 1903 does not show any well-marked correlation among ten-day temperatures, it will be of interest to trace out the cause of the seeming divergence. It would be better that this should be done by another; but some comparisons by the present writer may serve at least as suggestions on the subject.

We remark at the outset that there is no inherent necessity that the fluctuations in the seven regions selected by Langley should show any close relation with those of the three regions chosen in the present work. Such a relation can only be regarded as more or less probable according to circumstances.

The question now presents itself how far the seeming divergence arises from accidental fluctuations in the special data made use of, and how far to differences in the method of investigation. The methods of treatment are different in that the present work includes only regions of low or middle latitudes, while those chosen by Langley include northern regions also, especially Siberia. Thus a seeming discordance in the course of any one year is not surprising. I have not made a careful comparison of the two results except in the most striking case. The most important decade of comparison in the work is the first of 1903. The *Dekadenberichte* show an extraordinary rise of temperature during this term, while by reference to Table XII, 1903, of the present work, it will be found that the general mean deviation here found is only $+ 0°.5$. Considering this decade individually the evidence afforded by the *Dekadenberichte* is vastly more complete for the world at large. The positive departure was best marked in European Russia and Siberia, reaching its maximum at Orenburg, where the temperature was $12°.1$ C., or more than $20°$ Fahr. above the normal. But it covered the whole of Europe, Scandinavia excepted. Now, these regions I have mentioned are not included in the present work because the effect of any admissible change in the sun's heat on their temperature would be very slight through a ten-day term, especially in January. Although the general mean for the equatorial region is positive, it is not at all accented as in the wider range of regions used by Langley.

For our present purpose the important question is whether we can attribute this remarkable rise of temperature to an increase of the solar radiation. The reply is that, if there was such an increase during the decade in question, its effect would have been felt mainly in the equatorial regions, and but slightly in northern Europe and Siberia. We therefore conclude only that great fluctuations of temperature occur which we cannot attribute to changes in the sun's radiation, because they do not extend to the regions where such changes would have their greatest effect.

www.ingramcontent.com/pod-product-compliance
Lightning Source LLC
Chambersburg PA
CBHW081335190326
41458CB00018B/6004